Carbon and TMDs Nanomaterials for Energy Applications

CARBON AND TMDs NANOMATERIALS FOR ENERGY APPLICATIONS

Editor

Ashish Kumar Mishra

Indian Institute of Technology (BHU), Varanasi, India

World Scientific

NEW JERSEY · LONDON · SINGAPORE · BEIJING · SHANGHAI · HONG KONG · TAIPEI · CHENNAI · TOKYO

Published by

World Scientific Publishing Co. Pte. Ltd.

5 Toh Tuck Link, Singapore 596224

USA office: 27 Warren Street, Suite 401-402, Hackensack, NJ 07601

UK office: 57 Shelton Street, Covent Garden, London WC2H 9HE

British Library Cataloguing-in-Publication Data

A catalogue record for this book is available from the British Library.

CARBON AND TMDs NANOMATERIALS FOR ENERGY APPLICATIONS

ISBN 978-981-12-8339-0 (hardcover)
ISBN 978-981-12-8340-6 (ebook for institutions)
ISBN 978-981-12-8341-3 (ebook for individuals)

For any available supplementary material, please visit
https://www.worldscientific.com/worldscibooks/10.1142/13596#t=suppl

Typeset by Stallion Press
Email: enquiries@stallionpress.com

Dedicated to Her Holiness
Shree Mata Ji Nirmala Devi

Preface

The global energy demands have been continuously increasing in the last few decades. About 80% of this demand is currently fulfilled by nonrenewable fossil fuels, but their extensive utilization causes global warming, environment detrimental, ozone depletion, ocean acidification, and energy crisis. These issues have significant impact on social and economic development of every country, as well as on human, plant, and animal life. Therefore, the researchers have focused on the development of cost-effective and sustainable renewable energy technologies for energy generation and storage such as supercapacitors, metal–air batteries, lithium and sodium ion batteries, electrochemical hydrogen production, solar-driven steam generation, and fuel cell. To address these developments for upcoming energy technologies, carbon and transition metal dichalcogenide (TMD)-based nanomaterials have emerged as a promising candidate owing to their excellent electrochemical activity, high conductivity, large surface area, multifunctionality, nanodevice integration, mechanical flexibility, etc. To enhance the performances of existing energy devices, various modifications of these nanomaterials, such as heterostructure formation and doping with other atoms, have also received significant attention. In this regard, considerable efforts have been made to develop viable synthesis techniques to achieve different morphologies that affect the exotic physical properties of these nanomaterials, thus deciding the targeted energy applications.

This book solely focuses on the carbon and TMD nanomaterials for energy applications and deals with the technological importance of these well-established nanomaterials for their wide range of applicability in the area of energy generation and storage. Chapter 1 is devoted to Introduction (including electronic structure) and general applicability of carbon materials (graphene and carbon nanotubes (CNT)) and TMDs, while next two chapters (Chapters 2 and 3) discuss various synthesis and characterization approaches along with the key properties of carbon (graphene and CNTs) and TMD (MoS_2, $MoSe_2$, etc.) nanomaterials. The successive three chapters (Chapters 4-6) deal with utilization of these nanomaterials in various energy generation devices such as fuel cell, electrochemical hydrogen production, and solar-driven steam generation. Last three chapters (Chapters 7-9) survey the employment of these nanomaterials for energy-storage application such as supercapacitors, metal–air and metal-ion batteries.

This book aims to encourage the readers to know about the recent developments in the field of energy applications of carbon and TMD nanomaterials. In addition, this book will also encourage young scientists and researchers to further explore the properties and applications of carbon and TMD nanomaterials.

Dr. Ashish Kumar Mishra
Varanasi, India

Acknowledgement

First and foremost, I would like to thank the great GODDESS for giving me this human life and guiding me through all endeavors. I am indebted to my grandparents and parents for my upbringing through their care, sacrifice, and guidance. The completion of this book is possible with contributions from many individuals, scientists, and researchers. I would like to express my heartfelt gratitude and thanks to all the contributors of this book for providing quality chapters on recent trends on energy applications of carbon and TMD nanomaterials. I am thankful to my PhD. students (Ankita, Rohit, Prince, Jay, and Priyanka) for their support in editing this book. I owe my deep gratitude to Ms. Sandhya Devi, Editor, World Scientific Publishing Company and her team for providing this opportunity and helping in the completion of this book. I am also thankful to the Science and Engineering Research Board (SERB), DST, Government of India, and IIT (BHU) for providing support to conduct research in the field of energy materials. Last but not the least, I would like to thank my family-my mother Mrs. Pushpa Mishra, my wife Dr. Jaya Tripathi, and my son Master Revansh Mishra, for their valuable support and sacrifices during the preparation of the book.

Dr. Ashish Kumar Mishra
IIT BHU Varanasi, India

About the Editor

Dr. Ashish Kumar Mishra completed his M.Sc. in Physics from V.B.S. Purvanchal University, Jaunpur, in 2005 and Ph.D. in Physics from Indian Institute of Technology, Chennai, in 2011 with the *Best Ph.D. Thesis Award*. He worked as a Post-Doctoral Research Associate at Rensselaer Polytechnic Institute, USA (Jan 2012-Dec 2014) and as INSPIRE Faculty at IISER Bhopal, India (Feb 2015–Sep 2016) before joining IIT (BHU) in October 2016 as an Assistant Professor in School of Materials Science and Technology. He is working in the field of thermal transport study, optoelectronics, energy, and environment applications of carbon and TMD nanomaterials. He is an expert in the field of electrochemical device applications of 2D materials such as hydrogen production, supercapacitors, and batteries. He is also an expert in the field of optothermal Raman spectroscopy and SERS techniques. He has received prestigious national awards such as INYAS membership (2022), Early Career Research Award (2017), and INSPIRE Faculty Award (2014). He has published his research work in more than 50 high-impact international journals and contributed to multiple book chapters, and filed more than 8 national/international patents.

Homepage: https://iitbhu.ac.in/dept/mst/people/akmishramst
E-mail: akmishra.mst@iitbhu.ac.in

List of Contributors

Dr. Bishnu Pada Majee

Department of Materials Science and Engineering, Rensselaer Polytechnic Institute, Troy, NY 12180, USA; Department of Mechanical, Aerospace, and Nuclear Engineering, Rensselaer Polytechnic Institute, Troy, NY 12180, USA.

Ankita Singh

School of Materials Science and Technology, Indian Institute of Technology (Banaras Hindu University), Varanasi, India.

Dr. Ashish Kumar Mishra

School of Materials Science and Technology, Indian Institute of Technology (Banaras Hindu University), Varanasi, India.

Dr. Shanu Mishra

School of Materials Science and Technology, Indian Institute of Technology (Banaras Hindu University), Varanasi, India.

Akshaya S Nair

Department of Physics and Electronics, CHRIST (Deemed to be University), Hosur Road, Bengaluru 560029, India.

Soju Joseph
Department of Physics and Electronics, CHRIST (Deemed to be University), Hosur Road, Bengaluru 560029, India.

Dr. R. Imran Jafri
Department of Physics and Electronics, CHRIST (Deemed to be University), Hosur Road, Bengaluru 560029, India.

Rohit Kumar Gupta
School of Materials Science and Technology, Indian Institute of Technology (Banaras Hindu University), Varanasi, India.

Prince Kumar Maurya
School of Materials Science and Technology, Indian Institute of Technology (Banaras Hindu University), Varanasi, India.

Dr. Higgins Wilson
Department of Physics, Institute of Chemical Technology Mumbai, Nathalal Parekh Marg, Mumbai 400019, India.

Dr. Neetu Jha
Department of Physics, Institute of Chemical Technology Mumbai, Nathalal Parekh Marg, Mumbai 400019, India.

M. Manuraj
Chemical Sciences and Technology Division, CSIR-National Institute of Interdisciplinary Science and Technology (CSIR-NIIST), Thiruvananthapuram, Kerala 695019, India.

Visakh V Mohan
Department of Physics, University of Kerala, Kariavattom, Thiruvananthapuram, Kerala, 695581, India.

Dr. R. B. Rakhi
Materials Science and Technology Division, CSIR- National Institute of Interdisciplinary Science and Technology (CSIR-NIIST), Thiruvananthapuram, Kerala 695019, India.

Dr. Vimal K. Tiwari
Department of Physical Sciences, Banasthali Vidyapith, Banasthali Vidyapith, Banasthali 304022, India.

Dr. Rajendra Kumar Singh
Ionic Liquid and Solid-State Ionics Lab, Department of Physics, Institute of Science, Banaras Hindu University, Varanasi 221005, India.

Contents

https://doi.org/10.1142/9789811283406_0001

Chapter 1

An Overview of Carbon and Transition Metal Dichalcogenide Nanostructures

Bishnu Pada Majee[1,2], Ankita Singh[3], and
Ashish Kumar Mishra[3]

[1]*Department of Materials Science and Engineering,
Rensselaer Polytechnic Institute, Troy, NY 12180, USA*
[2]*Department of Mechanical, Aerospace, and Nuclear Engineering,
Rensselaer Polytechnic Institute, Troy, NY 12180, USA*
[3]*School of Materials Science and Technology, Indian Institute of
Technology (Banaras Hindu University), Varanasi 221005, India
Email: akmishra.mst@iitbhu.ac.in*

Abstract

Industrialization, technological advancement, and rapid population growth have placed immense pressure on energy resources. Rapidly diminishing fossil fuels and the unfavorable environmental impacts associated with their use have further emphasized the necessity for alternate energy sources. Researchers have been continuously seeking trustworthy, secure, and environmentally friendly energy sources, and found that carbon and two-dimensional (2D) materials such as transition metal dichalcogenides (TMDs) are suitable for energy generation (hydrogen production, fuel cells, and solar cells) and storage (supercapacitor and battery) applications. This chapter introduces the

1

basic structure of carbon-based materials (graphene and carbon nanotubes), TMD nanomaterials (mainly MoS_2 and $MoSe_2$), and their applications. Carbon and TMDs materials show various electronic properties, such as conducting, semiconducting, and metallic, depending on the atomic arrangement of the atoms in the lattice.

1.1 Introduction

In the 18th century, wood was the primary energy source followed by coal, after the invention of the steam engine. By 1970, oil and natural gas fossil fuels became the primary sources of energy in the world due to their high energy density and availability. However, in the 21st century, fossil fuels are considered to be unsustainable for the earth in the long run due to their rapid depletion and adverse effects on the environment. Many scientists and researchers are actively transitioning toward renewable energy sources to resolve these issues. Currently, energy generation and storage are the main challenges to the scientific community. Over the past few decades, researchers have developed renewable and sustainable energy sources and storage systems such as electrochemical cells for hydrogen production, supercapacitors, and batteries. These systems require efficient electrode materials with properties like good electrical conductivity, high surface area, and good catalytic activity. Carbon nanomaterials, such as graphene and carbon nanotubes (CNTs), and inorganic transition metal dichalcogenides (TMDs), such as MoS_2, WS_2, and $MoSe_2$, appear to be promising candidates for various energy applications due to their unique properties and exceptional performance. Graphene, the most familiar 2D material, has shown high electrical conductivity, mechanical strength, and thermal stability, making it suitable for energy storage and conversion devices. Its large surface area and excellent charge mobility enable efficient electrochemical energy storage in supercapacitors and batteries. CNTs and carbon-based composites have also shown remarkable catalytic activities for application in fuel cells, enhancing energy conversion efficiency. Other 2D nanomaterials, such as TMDs and black phosphorus, possess semiconducting properties and can be used in solar cells, photodetectors, along with energy storage systems. The exceptional properties of these materials offer immense potential for advancing energy technologies, paving the way for a sustainable and clean energy future. This chapter will introduce the basic structure of graphene, CNTs, and TMDs along with their applications in energy generation and storage devices.

1.2 Carbon Nanostructures

Carbon is an indispensable, inorganic, nonmetallic material having a vital place in materials science and technology. It is old but still relevant to the development of modern technologies due to various morphologies it can exist in. Carbon is an essential component for all living organisms and has various geometries, such as linear, planar, and tetrahedral bonding, to form different structures [Sisto *et al.*, 2016]. Carbon materials are widely used in industrial applications such as aerospace, chemical, and nuclear reactors [Bellucci *et al.*, 2007; Popa-Simil, 2008]. The most common allotropes of carbon are graphite and diamond, which are abundant in nature. Diamond is the hardest material due to sp^3 hybridization, whereas graphite is one of the softest materials due to sp^2 hybridization. This is mainly ascribed to the dissimilar atomic arrangement of carbon atoms in the lattice, which plays an important role in determining the physical properties of carbon.

Graphite is a three-dimensional (3D) allotrope of carbon with a hexagonal structure, where atoms are organized in a honeycomb network with a distance of ~0.142 nm, as shown in Fig 1.1(a). Graphite can be found in nature as flakes or lumps but rarely in a monocrystalline form. Graphene layers are stacked over each other and bonded by a weak force. Since the invention of the pencil in 1564, graphite has been widely used for writing. The schematic diagram of a diamond is shown in Fig 1.1(b). In diamond, the carbon atoms are bonded with each other via covalent bond, and each carbon atom is surrounded by four carbon atoms. The nanocrystalline forms of carbon include CNTs, fullerenes (buckyballs), and graphene. In 1991, Iijima first synthesized CNTs with an inner diameter of 4 nm [Iijima, 1991]. In 1996, Harry Kroto, Robert Curl, and Richard Smalley were awarded

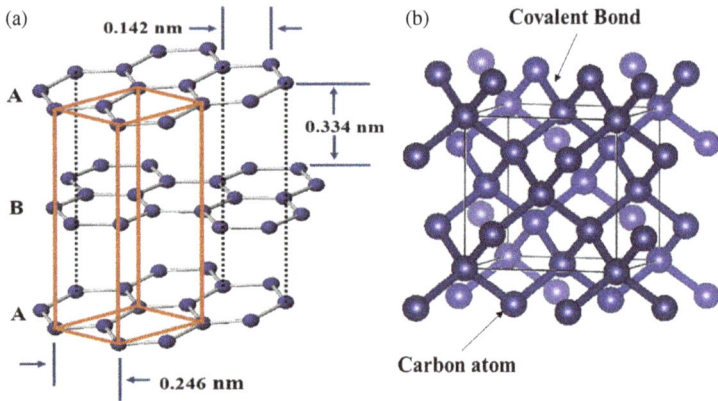

Fig 1.1   Schematic diagram of (a) hexagonal graphite and (b) diamond. Adapted from Liu & Zhou [2014].

the Noble Prize in the field of chemistry for the discovery of a noble molecule called Fullerene. Novoselov *et al.* [2004] first discovered the 2D carbon material called Graphene — a monolayer of carbon atoms arranged in a honeycomb lattice. The CNTs are rolled graphene structures having an axial symmetry and can be classified based on layer numbers like single-walled CNTs (SWCNTs), double-walled CNTs (DWCNTs), and multiwalled CNTs (MWCNTs). Graphene and SWCNTs possess outstanding electronic and mechanical properties from fundamental and application points of view. CNTs are 10,000 times thinner than human hair, whereas graphene is 300,000 times thinner than a sheet of paper. Generally, nanotube thickness varies from 1 to 100 nm depending on the presence of the layer (the thickness of single-layer graphene is ~3.4 Å) [Tan *et al.*, 2013]. CNTs have incredible mechanical strength and are the fastest known oscillators (>50 GHz). Depending upon the structure, that is, the arrangement of the carbon atoms, SWCNTs can be metallic or semiconducting [Cabria *et al.*, 2003]. Graphene and CNTs are promising crystalline carbon nanostructures for different technological applications such as supercapacitors, electronics, and sensors. The physical properties of CNTs mainly depend on the molecular structure and tube diameter. The electrical conductivity of metallic CNTs is ~10^3 times higher than copper, while the semiconducting nature of these CNTs varies with chiral pitch and diameter. The carbon materials show excellent thermal conductivity at room temperature, that is, 3,500 and 5,300 W m^{-1} K^{-1} for CNTs and graphene, respectively.

1.2.1 *Graphene*

Graphene consists of a single layer of carbon atoms that form graphite, as shown in Fig 1.2. It was isolated after 440 years of pencil invention [Novoselov *et al.*, 2004]. The name "graphene" is made up of the prefix "graph" from graphite and the "ene" from carbon–carbon double bonds [Bianco *et al.*, 2013] and it is the thinnest graphite layer with a thickness of 0.34 nm. Carbon atoms undergo sp^2 hybridization to produce graphene.

Graphene can be prepared by mechanical exfoliation from the bulk graphite due to weak bonding between the graphite layers. The properties of the resulting material depend on the stacking sequence. In naturally occurring graphite, the ABAB stacking sequence of layers is observed with an interlayer {002} spacing of ~0.34 nm, as shown in Fig 1.1. The unit cell of graphite is shown by the orange mark in

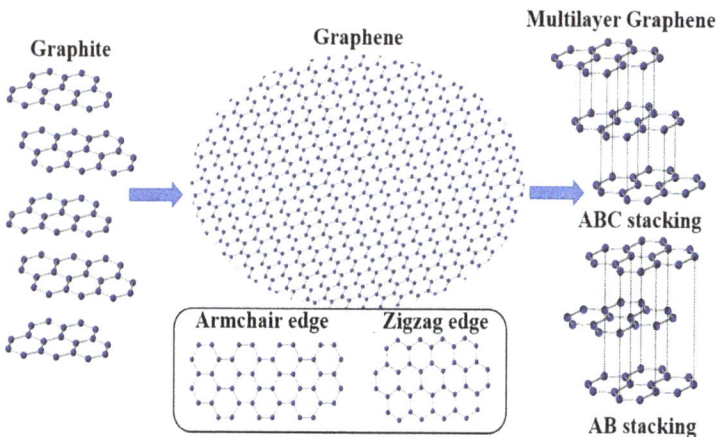

Fig 1.2 Schematics of graphite and graphene. The AB and ABC stacking sequences of graphene. Adapted from Meyer *et al.* [2007].

Fig 1.1 and it contains four carbon atoms. Its space group is P6$_3$/mmc (D$_{6h}$) [Zhou *et al.*, 2017]. In the case of graphite, it has been found that the value of the *c*-axis is nearly 0.67 nm. If more than one graphene layers are stacked, the stacking sequence may be AB or ABC, as shown in Fig 1.2. Graphene shows exceptionally high crystalline quality and outstanding electronic properties. It has been observed that the valence band (VB) and conduction band (CB) meet at the Dirac point in graphene. Due to linear dispersion in the Brillouin zone, graphene material has a zero effective mass, leading to interesting electronic and vibrational properties. Generally, the two layers of the graphene mound by AB stacking are named Bernal stacking arrangement. In AB stacking, the atom of the top layer is situated at the center of the hexagonal layer of graphene of the bottom layer, as shown in Fig 1.2.

1.2.2 *Carbon Nanotubes*

The structure of CNTs was first envisaged as a rolled-up sheet of graphene. The structures and symmetries of any material are essential to understand its physical behavior, as they are directly related to the atomic arrangement in the crystal. After the discovery of CNTs, it became obvious that an entirely new framework would be needed to explain this material. A free carbon atom possesses six electrons, with an electronic structure of $1s^2\ 2s^2\ 2p^2$. In CNTs and graphene, one of the 2s electrons goes to 2p to form the covalent bonds, and then the orbitals are hybridized. To understand CNTs, a basic idea about graphene is needed. The schematics of graphene with its stacking sequence forming armchair and zigzag edges are shown in Fig 1.2. The schematics of SWCNTs and MWCNTs are shown in Fig 1.3. CNTs are 1D carbon allotropes, having hollow, well-ordered graphitic nanomaterials made of cylinders of sp^2-hybridized carbon atoms with axial symmetry. The electronic properties of SWCNTs are mainly controlled via their chirality. There are different types of CNTs depending on the arrangement of carbon atoms. The unit cell is a part of CNT and the spiral symmetry of the nanotube is denoted by

SWCNT   MWCNT

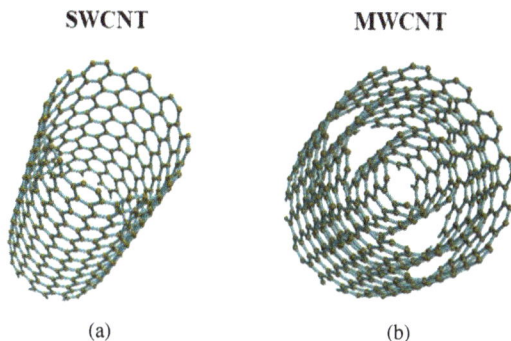

(a)   (b)

Fig 1.3   Schematics of (a) SWCNT and (b) MWCNT. Adapted from Wong & Akinwande [2010].

vectors. Nanotube structures can be represented by the following parameters: chiral vector (C_h), translation vector (T), and chiral angle (θ) [Ganesh, 2013].

The chiral and translational vectors can be denoted by $C_h = na_1 + na_2 \equiv (n, m)$ and $\mathbf{T} = t_1 a_1 + t_2 a_2 \equiv (t_1, t_2)$, respectively, where a_1 and a_2 are the base vectors, $t_1 = (2m + n)/dR$, $t_2 = -(2n + m)/dR$, $dR =$ gcd $(2n + m, 2m + n)$, and n and m are the lengths of chiral vectors. Other important parameters are chiral angle and chiral length, represented as $\cos \theta = (2n + m)/(2 \times (n^2 + m^2 + n \times m)^{1/2})$ and $L = a \times (n^2 + m^2 + n \times m)^{1/2}$, respectively, where a is the lattice constant. The diameter of the nanotube and the number of hexagons in the unit cell are represented as $dt = L/\pi$ and $N = (2 \times (n^2 + m^2 + n \times m)/dR)$, respectively. Fig 1.4 shows the hexagonal lattice of graphene with lattice indices of translations (n, m) with base vectors a_1 and a_2. The CNTs are formed by rolling about a translational vector ($\mathbf{T}$), as shown in Fig 1.4, and can be classified based on chiral vector and the corresponding chiral angle. There are three types of CNTs: zigzag ($m = 0$, $\theta = 0°$), armchair ($n = m$, $\theta = 30°$), and chiral ($n = m$, $0° < \theta < 30°$) [Sisto et al., 2016].

The electronic properties of SWCNT are directly related to the chiral vector. These three structures of CNTs possess different electronic properties. If the difference between n and m is a multiple of 3,

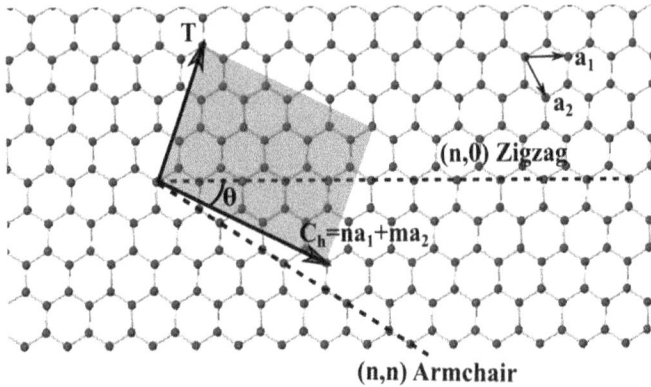

Fig 1.4   Hexagonal lattice of graphene with basis vector, translation vector, chiral vector, zigzag, and armchair. Adapted from Baghdadi *et al.* [2015].

it will be metallic in nature, else semiconducting [Charlier, 2002]. The armchair and chiral structures possess metallic properties whereas zigzag CNTs show semiconducting properties [Ganesh, 2013].

1.3 Transition Metal Dichalcogenides

The outstanding properties of graphene lead researchers' interest in other 2D materials, such as TMDs, mainly when brought down to the atomic scale. The composition of atoms and the thickness of TMDs layer play vital roles in determining the fundamental properties such as bandgap, surface-to-volume ratio, and catalytic activity. [Mas-Ballesté *et al.*, 2011]. Graphene is the thinnest 2D material with zero bandgap, whereas TMDs do not have this problem. The TMDs have a direct bandgap at monolayer, tunable bandgap (with temperature and layer number), good electrical conductivity, high absorption coefficient, high stability at environmental conditions (also in acidic and basic medium), etc. These materials can be metals, semimetals, insulators, and semiconductors in nature, depending on the atomic arrangement of transition metal and chalcogen atoms. Due to their weak interlayer interactions, it is possible to exfoliate these TMDs from their bulk counterpart and use them in optoelectronic devices, supercapacitors,

MX$_2$
M= Transition metal
X= Chalcogen

H																	He
Li	Be											B	C	N	O	F	Ne
Na	Mg	3	4	5	6	7	8	9	10	11	12	Al	Si	P	S	Cl	Ar
K	Ca	Sc	Ti	V	Cr	Mn	Fe	Co	Ni	Cu	Zn	Ga	Ge	As	Se	Br	Kr
Rb	Sr	Y	Zr	Nb	Mo	Tc	Ru	Rh	Pd	Ag	Cd	In	Sn	Sb	Te	I	Xe
Cs	Ba	La-Lu	Hf	Ta	W	Re	Os	Ir	Pt	Au	Hg	Tl	Pb	Bi	Po	At	Rn
Fr	Ra	Ac-Lr	Rf	Db	Sg	Bh	Hs	Mt	Ds	Rg	Cn	Uut	Fl	Uup	Lv	Uus	Uuo

Fig 1.5   Periodic table showing the transition metals and chalcogen atoms, highlighted in gray and sky blue colors, respectively. Adapted from Chhowalla *et al.* [2013].

hydrogen generation, batteries, lubricants, etc. The family of 2D materials encompasses a wide selection of compositions, including nearly all the periodic table elements. In the 2D family, TMDs are the most promising materials for next-generation energy and optoelectronic applications due to their outstanding optical, electrical, and mechanical properties at the nanoscale. TMDs are shown in Fig 1.5, where the metal atoms are in gray and the chalcogen atoms in sky blue. About 40 different layered TMDs are found in nature. Layered TMDs such as MoS_2, WS_2, $MoSe_2$, and WSe_2 can be synthesized via different top-down and bottom-up approaches [Singh *et al.*, 2023].

The most interesting properties of TMDs emerge when we go from bulk to few layers, showing quantum confinement. Its inversion symmetry disappears, which causes distinct changes in the properties of TMDs [Chhowalla *et al.*, 2013]. Thus, these 2D materials can play a fundamental role in next-generation device applications. The variable bandgap, high surface area, and high charge carrier mobility of the 2D TMDs pave the way for potential applications in optoelectronic, catalysts, energy storage, energy generation, and the field of sensing [Taylor *et al.*, 2014].

The general formula for 2D TMDs is MX_2, where M represents the transition metal and the X represents the chalcogen atoms.

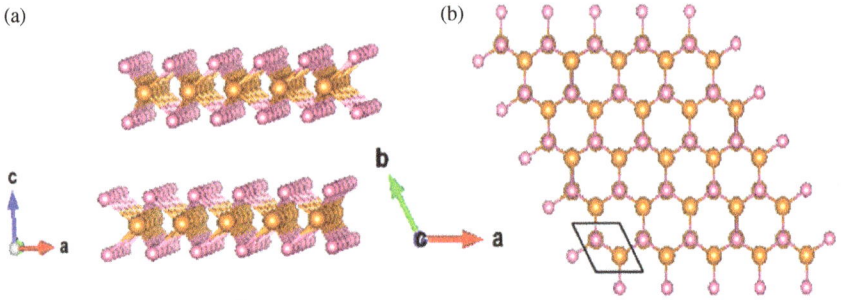

Fig 1.6   Schematic representation of (a) side view and (b) top view of the 2H-MoS$_2$ structure. Adapted from Jayabal *et al.* [2017].

The metal elements belong to groups 4, 5, and 6 in the periodic table, as shown in Fig 1.5, and the chalcogen atoms are S, Se, and Te. These materials make different 2D-layered TMDs of the form X-M-X, and the metal atom is sandwiched between two chalcogenide atoms [Chhowalla *et al.*, 2013]. The schematic representation of the side view of MX$_2$ material is shown in Fig 1.6(a) and the top view of MX$_2$ is illustrated in Fig 1.6(b). In the case of 2D TMDs, the adjacent layers are weakly coupled to produce the bulk crystal of a different polytype. The most remarkable properties of 2D TMDs are layer-dependent properties that have recently attracted the scientific community. It has been experimentally observed that the thickness of each layer in TMDs is ~6–7 Å. In this type of structure, intra-layer bonds are covalent but the van der Waal forces couple the layers of TMDs. The covalent bonds are much stronger compared to the van der Waal forces. This weak force permits the cleavage of bulk TMDs along the layer surface to form a few/monolayer material via exfoliation. In TMDs, the oxidation state of the transition metal (M) is +4 and for the chalcogen (X) atom is –2 [Chhowalla *et al.*, 2013]. The nonappearance of dangling bonds makes these materials more stable and long-lasting in different environmental conditions. The distance between the M–X bonds varies in the range of 3.15–4.03 Å, depending on the size of the M atoms and X ions. Many semiconducting TMDs show an indirect bandgap in the bulk state and transit to a direct bandgap in the case of a monolayer. For example, bulk MoS_2 possesses an indirect bandgap of 0.99 eV and a direct bandgap of 1.74 eV in monolayer [Singh *et al.*, 2023].

TMDs are typically composed of a transition metal paired with a chalcogen, similar to the layered structure of graphite. Each metal M atom is sixfold coordinated whereas the X atoms form either trigonal prismatic or octahedral coordination. The three chalcogen atoms are placed in a trigonal prismatic arrangement; however, in an octahedral arrangement, they are staked as shown in Fig 1.7. The most widely studied TMD in group 6 is MoS_2. The bulk MoS_2 (or $MoSe_2$) shows polymorphism depending on the atomic arrangement. The schematic diagrams of three distinct phases of MoS_2 are 1T, 2H, and 3R, as shown in Fig 1.7. The numbers 1, 2, and 3 represent the layer numbers present in the unit cell and the capital letters represent the symmetry, that is, T for trigonal, H for hexagonal, and R for rhombohedral. Among these polymorphs of MoS_2, 2H is the most stable. The 2H- and 3R-MoS_2 have trigonal prismatic coordination whereas the Mo atom is octahedrally coordinated in 1T-MoS_2 [Wang *et al.*, 2012]. However, the dissimilar arrangement of layers makes the two

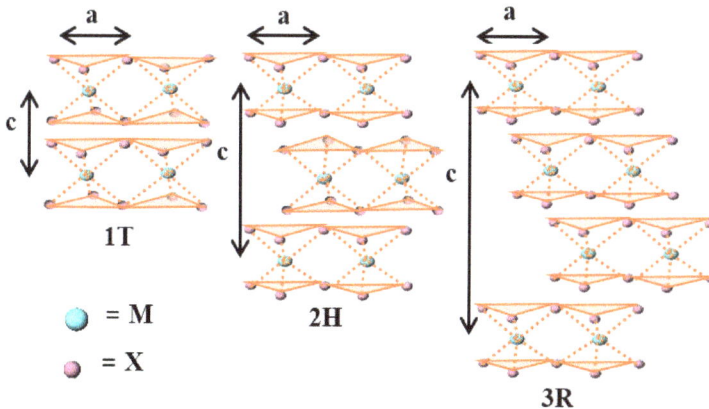

Fig 1.7 Schematics of a TMD: structural polytypes: 1T (tetragonal symmetry), 2H (hexagonal symmetry), and 3R (rhombohedral symmetry) phases of MX_2 material. Adapted from Wang *et al.* [2012].

semiconducting polymorphs of 2H and 3R-MoS$_2$ structurally different. The 3R and 1T-MoS$_2$ can be effortlessly converted into 2H polytype via heating, strain engineering, electron doping, etc. [Duerloo *et al.*, 2014]. The stacking sequence in the 2H phase of MoS$_2$ is AbA BaB (capital letters denote the chalcogen atoms and the lower-case letters denote metal atoms). In the synthetic 3R MoS$_2$ phase, the stacking sequence is AbA CaC BcB. In both phases, 2H and 3R, the metal coordination is trigonal prismatic as shown in Fig 1.7 [Cho *et al.*, 2015]. In TMDs, the electronic structure is highly dependent on the atomic positions of the M atom and its *d*-electron count. [Chhowalla *et al.*, 2013].

The sulfide- and selenide-based TMDs have a stable 2H phase, whereas their 1T phase is metastable. For tellurides, the 1T phase is more favorable. Conversion of 2H into 1T is possible under certain conditions, such as Li-ion intercalation, strain, electric field, and heating. The 2D TMDs have occupied an appreciable position among the layered materials for different applications because of their exceptional properties such as stability, high absorption coefficient, good mobility, tunable bandgap, etc.

1.4 Applications of Graphene, CNTs, and TMDs

The economic development depends on the technological advances made by the researchers. In the last decades, energy production, harvesting, and storage fields have remained one of the most critical expenses. Renewable and zero-emission technologies have garnered great attention from researchers due to global warming resulting from the combustion of fossil fuels and its impact on the environment. The carbon and TMD nanomaterials show promising physical and chemical properties, such as excellent electrical conductivity, chemical stability, and catalytic activity, which can be applied in various fields, as shown schematically in Fig 1.8.

Among carbon nanostructures, CNTs (SWCNTs, DWCNTs, and MWCNTs) and graphene are usually used as electrode materials in energy and electronic devices due to their unique porous nature, high mechanical strength (~1 TPa), good stability in acidic/basic

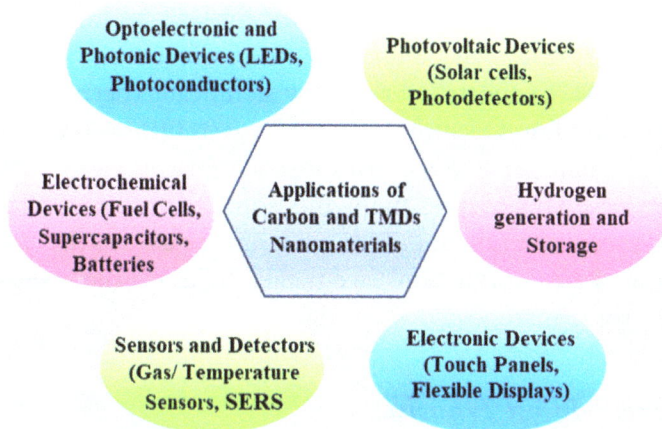

Fig 1.8   Schematics showing different applications of 2D materials.

environments, high electrical conductivity ($\sim 10^6$–10^7 S m^{-1}), good thermal conductivity (~ 3000 W m^{-1} K^{-1}), etc. The SWCNT is theoretically expected to have a high surface area of ~ 1300 m^2 g^{-1}, but experimentally it is in the range of 100–1000 m^2 g^{-1}. Among the 2D family of materials, the first used material for technological applications is the 2D carbon material, that is graphene, which shows an even higher theoretical surface area of 2630 m^2 g^{-1} along with good mechanical strength (~ 1 TPa), greater thermal conductivity (~ 5000 W m^{-1} K^{-1}), high electrical conductivity ($\sim 10^8$ S m^{-1}), and high electronic mobility ($\sim 200,000$ cm^2 V^{-1} s^{-1}). However, due to the zero bandgap of graphene, it shows several shortcomings, and hence its application in future semiconducting devices is very challenging. The scientific community has tried to develop graphene-like 2D materials with built-in semiconducting properties like intrinsic bandgap. The chemistry of MX_2 materials gives distinct opportunities for going beyond graphene. It opens up new technological and fundamental paths for inorganic 2D materials. Due to the presence of intrinsic bandgap in MoS_2 (indirect ~ 1.2 eV for bulk and direct ~ 1.8 eV for monolayer), $MoSe_2$ (indirect ~ 1.1 eV for bulk and direct ~ 1.5 eV for monolayer), WS_2 (indirect ~1.3 eV for bulk and direct ~2.0 eV for monolayer), and WSe_2 (indirect ~1.2 eV for bulk and direct ~1.7 eV for monolayer), they can be used in semiconductor devices. In addition, these TMD nanostructures show significant thermal (~20–150 W m^{-1} K^{-1}) and good electrical conductivity (0.001–0.1 S m^{-1}) which also makes them suitable for electrode material in energy and electronic devices. This chapter highlights some of the practical applications of these carbon and TMD nanomaterials in the following subsections. Later in this book, we mainly focus on the energy applications of these materials.

1.4.1 *Field-Effect Transistors*

Field-effect transistor (FET) is an important integral in modern electronic devices. A typical FET consists of source and drain regions, connected via a thin area called channel. It is a type of transistor in which the electric field is used to control the channel conductivity. 2D carbon materials such as graphene have very high mobility and thermal conductivity, but their zero bandgap makes them unsuitable for FET applications. However, the 1D semiconducting CNTs can be used for fabricating both *n*-type and *p*-type channel-based FET devices. Being 1D, scattering probability is greatly reduced as there is no boundary scattering and conduction occurs along its surface where all the bonds are stable and saturated. Recently, TMDs such as MoS_2 and WS_2 have been used as channel materials between the electrodes in FET due to their layer-dependent bandgap, flexible nature, and good conductivity. Researchers have shown that the performance of TMD-based FET is better than the traditional silicon-based transistors due to the natural bandgap, thin nature, and layer-dependent charge transport. Liu *et al.* [2020] used CVD-grown monolayer MoS_2 in FET and found a high on/off ratio of 10^8 and carrier mobility up to 118 cm^2 V^{-1} s^{-1}. Moreover, heterostructures formed by stacking different 2D materials can further enhance the functionalities of FETs. Wu *et al.* [2016] demonstrated the growth of WS_2/MoS_2 heterostructure via CVD method and showed the enhanced performances of the heterostructure-based FET device compared to MoS_2

layer, due to additional electron injection from WS_2 under thermal equilibrium.

1.4.2 *Photodetector*

The photodetector is an optoelectronic device that converts absorbed photons into electrical signals. The high-responsivity photodetectors have tremendous societal importance in different real-life applications such as sensing, night vision camera, missile warning, and biomedical imaging. Graphene has been widely investigated in advanced photodetectors as it has several desired properties such as high carrier mobility, broadband absorption, and flexibility. The extremely high carrier mobility rules the response time showing that graphene-based photodetectors are capable of ultrafast operation [Long *et al.*, 2019]. Graphene is a gapless semimetal that absorbs ultraviolet, visible, infrared, and terahertz frequency ranges [Li *et al.*, 2017]. However, there are some big challenges for developing graphene-based photodetectors due to their intrinsic properties such as zero bandgap, low absorption coefficient, and ultrafast carrier recombination (picoseconds). Thus, the intrinsic graphene-based photodetector gives a very low value of photoresponsivity. Therefore, high photoresponsivity and on/off ratio materials are required for photodetector applications. The group VI-TMDs materials such as MoS_2 and WS_2 are semiconducting and their bandgap varies with layer number. The CVD-grown MoS_2 over the Si substrate is generally *n*-type and makes a *p–n* junction with the *p*-type Si substrate. Majee *et al.* [2019] showed the photoresponsivity of 0.1413 A W^{-1} at –2 V for pristine few-layer MoS_2/Si under white light illumination (0.15 mW cm^{-2}). The morphology of MoS_2 affects the photoresponsivity of the device. A very high responsivity of ~7.37 A W^{-1} at –2 V has been observed for vertically oriented MoS_2 nanosheets/Si-based photodetectors [Majee *et al.*, 2020]. Zheng *et al.* [2016] demonstrated the flexible and highly stable ultra-broadband photodetector using multilayer WSe_2 films, synthesized via pulsed-laser deposition technique. They showed the reversible photoresponsivity of 0.92 A W^{-1}, a fast response time of 0.9 s, spectral sensitivity in the range of 370–1064 nm, and a stable photocurrent up to a bending radius of 10 mm. Selamneni *et al.* [2020] demonstrated the cost-effective, biodegradable, and flexible near-infrared photodetector using $MoSe_2$ nanoflowers on cellulose paper, synthesized via vacuum filtration method. The fabricated photodetector showed a responsivity of 9.73 mA W^{-1}. Li *et al.* [2020] fabricated the flexible broadband photodetector using WS_2 nanosheets, synthesized via hydrothermal interaction and vacuum filtration method. This device showed the responsivity of 4.04 mA W^{-1} and responded to a broadband wavelength of 532–1064 nm. Thus, it is observed that the TMD nanostructure films show excellent photoresponsivity in the visible to IR range.

1.4.3 *Surface-Enhanced Raman Spectroscopy*

In the last two decades, scientists have used the surface-enhanced Raman spectroscopy (SERS) technique for quality control in medical science, the food industry, and laboratory. The SERS phenomenon occurs due to electromagnetic and chemical enhancement. Electromagnetic enhancement occurs mainly in metal nanoparticles. In the case of semiconducting materials, the enhancement occurs due to a chemical mechanism. The chemical enhancement depends upon the energy level of the SERS substrate and analyte molecules. Researchers have mostly used novel materials such as gold, silver, and copper in SERS detection. However, in the last decade, carbon and TMD nanomaterials have been identified as cost-effective SERS substrates (alternative to expensive metals) due to their unique electronic properties, high surface area, high stability in harsh environments, and surface roughness. Majee *et al.* [2020] showed the sub-nanomolar (10^{-10} M) detection limit for organic dyes (R6G and methyl orange) using CVD-grown, vertically oriented few-layered MoS_2 nanostructures as SERS substrates, owing to enhanced light trapping and effective dye adsorption over the synthesized morphology. Ghopry *et al.* [2019] synthesized TMD (MoS_2 and WS_2) nanodomes/graphene heterostructures and obtained the detection of 10^{-11} M to 10^{-12} M for R6G, higher than individual MoS_2, WS_2, or graphene. This study suggests that carbon/TMDs heterostructure can be used to improve the SERS signals for molecular detection.

1.4.4 *Gas Sensors*

Gas-sensing devices have received enormous interest for the detection of various organic and inorganic (NO_2, NH_3, CO, H_2S, etc.) environmental pollutants in real life. Conventional gas sensors require a change in electrical properties of materials for detecting various gasses, but their utilization is limited by operating temperature and stability. Though conducting polymer-based gas sensors can be operated at room temperature, their electrical properties get degraded by humidity. Nowadays, researchers are trying to explore sustainable gas sensors that are fully reversible at room temperature because a room-temperature gas sensor consumes less power, as the desorption of gas does not require thermal energy. 2D nanomaterials such as graphene, MoS_2, WS_2, $MoSe_2$, WSe_2, and $ReSe_2$ can be employed as active materials in gas sensors, owing to their high surface areas and unique semiconducting features with tunable bandgaps. The large area growth of 2D materials on plastic and polymeric substrates promotes the advancement of cost-effective and mechanically stable gas-sensing devices. Kim *et al.* [2015] proposed a graphene-based sensor that can detect NO_2 gas without external heating and found that by increasing the bias voltage, the response and recovery were improved. Singh *et al.* [2020] developed a highly sensitive ammonia (NH_3) gas sensor based on $MoSe_2$ nanosheets, synthesized via liquid exfoliation method. The sensor showed good sensitivity (5.5%) down to 1 ppm with a fast response and recovery time of 15 and 135 s, respectively. They also showed the sensing mechanism that includes the adsorption kinetics and charge transfer between adsorbed NH_3 and MoS_2 via DFT simulation.

1.4.5 *Solar Cell*

In the last two decades, silicon (Si) has been widely used for photovoltaic applications due to its low bandgap, high stability, and abundant in nature. Nowadays, 80% of the world's energy supply comes from fossil fuels, however, the photovoltaic energy contribution is less than 0.04% [Wigley & Raper, 2001]. The TMDs possess some exciting properties such as ultrathin, highly transparent, lightweight, flexible nature, and bandgap in the visible to near-infrared range, which makes them suitable for photovoltaic devices. In a solar cell device, the electric field is formed due to the difference in the work function of the dissimilar materials at the junction. The electrons and holes are generated within the materials due to light absorption and a built-in electric field that separates these electrons at the junction. Researchers are making many attempts to improve its efficiency, either by doping, applying an interfacial layer, or antireflection coating. Doping increases the carrier concentration, the interfacial layer suppresses the recombination, and antireflection enhances the photon absorption by trapping the light. Song *et al.* [2015] fabricated graphene-silicon Schottky barrier solar cells. They achieved a power conversion efficiency of 15.6% considering all of them. Xu *et al.* [2019] prepared MoS_2/Si-based solar cells and showed that the conversion efficiency of ITO/MoS_2/p-Si/Ag (4.6%) was significantly enhanced from 1.1% for ITO/p-Si/Ag. The performance was enhanced due to the insertion of MoS_2 film that decreased the interfacial defects and increased the width of the depletion region in the solar cell.

1.4.6 *Photocatalysts*

The most abundant and clean energy source is solar energy. The solar energy that hits our earth in one hour has relatively more energy than humans use in a year. Thus, in recent times, researchers have worked extensively to harvest solar energy and use it for green energy generation. This is the most efficient technique to convert solar energy to chemical energy or solar fuels. It is the most promising and effective long-term solution to energy and environmental problems. Photons having energy greater than the bandgap of the semiconductors, produce photogenerated electrons in the CB and holes in the VB

[Luo *et al.*, 2016]. Renewable solar energy is used in photocatalysis to activate the chemical reaction at the semiconductor's surface. During that time, two reactions occur: one is oxidation from the photoinduced positive holes and the other is reduction from the photoinduced negative electron. In the last decade, 2D materials have shown great potential in photocatalytic applications because of their layer-dependent properties. Majee *et al.* [2019] demonstrate the photocatalyst application of few-layer MoS_2 for the degradation of R6G dye under a visible light intensity of 10 mW cm^{-2} for different periods. Wang *et al.* [2017] synthesized flower-like MoS_2 microspheres using hydrothermal methods with different pH values (9, 7, 5, 2, 1, 0.5, and 0.1). They observed that by lowering the pH value, there was a size reduction in the MoS_2 microflower that affected its photocatalytic activity. The MoS_2 microflowers synthesized at pH = 1 showed superior photocatalytic performances for the degradation of rhodamine-B and methylene blue under natural sunlight irradiation.

1.4.7 *Supercapacitor*

Supercapacitors or electrochemical capacitors play an important role in multiple energy devices with their excellent performance in energy and power densities, fast charge–discharge rate, and good stability. This device is a high-power device and it can work at high charge and discharge rates over several cycles. The electrode materials should have a large area and high electrical conductivity for the high capacitance of the device. Reduced graphene oxide (rGO) and CNTs show good capacitance behavior due to their porous nature and good conductivity. A specific capacitance of up to 276 F g^{-1} has been recorded for solvothermally prepared rGO [Lin *et al.*, 2011]. The 2D TMDs possess multilayers structure with significant surface area, which permits the intercalation of ions between the layers, providing countless potential for energy applications [Da Silveira Firmiano *et al.*, 2014]. The metallic phase TMDs show better capacitive performance compared to semiconducting ones due to better electron transport [Yang *et al.*, 2013].

1.4.8 *Batteries*

The development of reliable, renewable, and clean energy supplies has been viewed as a critical solution to address the energy issue and environmental deterioration in modern society. Unquestionably, rechargeable batteries are one of the greatest options for chemical energy storage, and the inherent properties of electrode materials are key to comprehending battery chemistry and enhancing battery performance [Meng *et al.*, 2017]. Carbon-based materials have unique properties such as large surface area, excellent electrical conductivity, highest surface-to-volume ratio, and better stability in harsh environments, which are promising in battery applications [Velraj *et al.*, 2015]. Similarly, TMDs such as MoS_2, $MoSe_2$, $ReSe_2$, and WS_2 have been regarded as other attractive electrode materials in energy storage devices. TMDs possess unique hexagonal crystal structures and chemical stability, which may help in greater cycling performance of TMDs electrodes. The TMDs are suitable electrocatalysts because of their layered structure, wide interlayer spacing, and large surface area, which is viable for ion diffusion and intercalation [Liu, 2022].

1.4.9 *Fuel Cell*

A fuel cell is a promising way to fulfill the energy demands of society with zero pollution. The operation of the fuel cell (energy conversion device) involves the hydrogen fuel that is continuously supplied at the anode and oxidant (oxygen from the air) that is continuously fed at the cathode. The problems faced by the fuel cell are their short life span, low power density per unit volume, and less durability due to sluggish oxygen reduction reaction kinetics that is occurring at the cathode. Carbon and TMD nanostructures show tremendous potential as electrocatalysts for fuel cells due to their remarkable physical properties, availability, processability, environmental friendliness, and relative stability in both acidic and basic mediums. Yun *et al.* [2011] fabricated Pt-graphene/MWCNT composite using spray coating from a Pt-graphene and MWCNT dispersion and observed their electrochemical performance in proton exchange membrane fuel cell

(PEMFC). The highly porous structure of MWCNT provided an electrical pathway for ORR and reduced the charge transfer resistance of the Pt-graphene/MWCNT composite cathode. They observed a four times increment in power density of the Pt-graphene/MWCNT composite, as compared to the Pt-graphene cathode. Hu and Chua [2016] synthesized MoS_2 fin-like nanostructures on carbon nanospheres in PEMFC and showed better electrochemical activity and stability of Pt/0.001 mg cm^{-2} MoS_2/carbon nanosphere, in comparison to Pt directly deposited on carbon nanostructure. The excellent catalytic performance was attributed to good water management of MoS_2 nanofilm.

1.4.10 *Hydrogen Evolution Reaction*

In light of recent environmental challenges, the scientific community is encouraged to develop sustainable and renewable energy devices to meet the growing demand for energy supplies. Modern research focuses on exploring the earth's abundant and low-cost materials that can replace the costly electrocatalysts like Pt for hydrogen evolution reaction (HER) [He *et al.*, 2016]. Electrochemical water splitting is regarded as the most promising method for clean and green hydrogen production with zero CO_2 emission. Carbon and TMD nanomaterials are used as highly efficient, inexpensive, and durable electrocatalysts for HER because of their unique electronic structure, large assessable surface area, maximum number of active edge sites, and ease of intercalation, which make them good candidates for energy generation and storage applications. The 1T MoS_2 has been studied more than the 2H phase as an electrocatalyst for HER, which supported by theoretical and experimental research. It was theoretically shown that the free energy (ΔG) of hydrogen adsorption on the MoS_2 edge is indeed close to thermo-neutral, indicating HER activity of MoS_2. The biggest problem in any electrochemical study is stability. Experimentally, the group-VI TMDs have confirmed decent stability in HER. It shows over 10,000 cycles with a minute increase in onset potential or the Tafel slope [Hinnemann *et al.*, 2005].

1.4.11 *Hydrogen Storage*

The onboard hydrogen energy vehicles require safe, light, compact, and affordable hydrogen. Carbon- and TMD-based materials are used as promising candidates for hydrogen storage because of their unique properties such as high porosity, layered structure, multilayer adsorption, tenability, and large accessible surface area. The use of these materials reduces the operational pressures while maintaining the high storage capacities of hydrogen, thus offering an alternative solution to conventional technologies. Chen *et al.* [2020] investigated the adsorption of hydrogen molecules on monolayer 1T and 1T' MoS_2 via first-principles study. They showed a better performance of hydrogen adsorption in 1T' MoS_2, compared to other doped or decorated 2H phase of MoS_2. They found the adsorption energy for hydrogen molecules for 1T' phase in the range of −0.2 to −0.6 eV with adsorption quantity up to 3.9 wt%. Yadav *et al.* [2020] reported enhanced hydrogen storage at room temperature in preheated self-aligned graphene oxide (GO) samples. The synthesized GO samples were preheated at three different temperatures (25°C, 250°C, and 400°C) and observed the hydrogen storage of ~1.5, 2.0, and 2.5 wt% at ~20 bar, respectively. The enhanced hydrogen storage capacity of preheated GO samples is due to self-aligned structure, topological defects, and increased hydrogen binding energy that helps in more hydrogen adsorption.

1.5 Conclusion

The above discussion is intended to summarize the structure and applications of carbon (CNT and graphene) and TMD nanostructure materials. The extraordinary properties of these materials make them suitable for various applications. The detailed synthesis process and properties of carbon (graphene and CNT) and TMDs (sulfides and selenides) such as MoS_2, $MoSe_2$, WS_2, and WSe_2 will be discussed in Chapters 2 and 3, respectively. Their detailed applications in various energy generation and storage devices such as fuel cells, electrochemical hydrogen production, solar-driven steam generation, supercapacitors, metal–air batteries, and lithium and sodium ion battery are respectively discussed from Chapter 4 to Chapter 9.

References

Baghdadi, H., Tounsi, A., Zidour, M. & Benzair, A. (2015). Thermal effect on vibration characteristics of armchair and zigzag single-walled carbon nanotubes using nonlocal parabolic beam theory. *Fuller. Nanotub. Carbon Nanostruct.* 23, pp. 266–272.

Bellucci, S., Balasubramanian, C., Micciulla, F. & Rinaldi, G. (2007). CNT composites for aerospace applications. *J. Exp. Nanosci.* 2, pp. 193–206.

Bianco, A., Cheng, H. M., Enoki, T., Gogotsi, Y., Hurt, R. H., Koratkar, N., Kyotani, T., Monthioux, M., Park, C. R., Tascon, J. M. D. & Zhang, J. (2013). All in the graphene family-A recommended nomenclature for two-dimensional carbon materials. *Carbon* 65, pp. 1–6.

Cabria, I., Mintmire, J. W. & White, C. T. (2003). Metallic and semiconducting narrow carbon nanotubes. *Phys. Rev. B-Condens. Matter Mater. Phys.* 67, p. 4.

Charlier, J. C. (2002). Defects in carbon nanotubes. *Acc. Chem. Res.* 35, pp. 1063–1069.

Chen, J., Cao, J., Zhou, J., Zhang, Y., Li, M., Wang, W. & Liu, X. (2020). Mechanism of highly enhanced hydrogen storage by two-dimensional 1T′ MoS_2. *Phys. Chem. Chem. Phys.* 22(2), pp. 430–436.

Chhowalla, M., Shin, H. S., Eda, G., Li, L., Loh, K. P. & Zhang, H. (2013). The chemistry of two-dimensional layered transition metal dichalcogenide nanosheets. *Nat. Chem.* 5, pp. 263–275.

Cho, S., Kim, S., Kim, J. H., Zhao, J., Seok, J., Keum, D. H., Baik, J., Choe, D.-H., Chang, K. J., Suenaga, K., Kim, S. W., Lee, Y. H. & Yang, H. (2015). Phase patterning for ohmic homojunction contact in $MoTe_2$. *Science* 349, pp. 625–628.

Da Silveira Firmiano, E. G., Rabelo, A. C., Dalmaschio, C. J., Pinheiro, A. N., Pereira, E. C., Schreiner, W. H. & Leite, E. R. (2014). Supercapacitor electrodes obtained by directly bonding 2D MoS_2 on reduced graphene oxide. *Adv. Energy Mater.* 4, pp. 1–8.

Duerloo, K. A. N., Li, Y. & Reed, E. J. (2014). Structural phase transitions in two-dimensional Mo-and W-dichalcogenide monolayers. *Nat. Commun.* 5, pp. 1–9.

Ganesh, E. N. (2013). Single walled and multi walled carbon nanotube structure. *Synth. Appl.* 2, pp. 311–320.

Ghopry, S. A., Alamri, M. A., Goul, R., Sakidja, R. & Wu, J. Z. (2019). Extraordinary sensitivity of surface-enhanced raman spectroscopy of molecules on MoS$_2$ (WS$_2$) nanodomes/graphene van der waals hetero-structure substrates. *Adv. Opt. Mater.* 7(8), p. 1801249.

He, Z. & Today, W.Q.-A.M. (2016). Molybdenum disulfide nanomaterials: structures, properties, synthesis and recent progress on hydrogen evolution reaction. *Appl. Mater. Today* 3, pp. 23–56.

Hinnemann, B., Moses, P. G., Bonde, J., Jørgensen, K. P., Nielsen, J. H., Horch, S., Chorkendorff, I. & Nørskov, J. K. (2005). Biomimetic hydrogen evolution: MoS$_2$ nanoparticles as catalyst for hydrogen evolution. *J. Am. Chem. Soc.* 127, pp. 5308–5309.

Hu, Y. & Chua, D. H. (2016). Synthesizing 2D MoS$_2$ nanofins on carbon nanospheres as catalyst support for proton exchange membrane fuel cells. *Sci. Rep.* 6(1), pp. 1–10.

Iijima, S. (1991). Helical microtubules of graphitic carbon. *Nature* 354, pp. 56–58.

Jayabal, S., Saranya, G., Wu, J., Liu, Y., Geng, D. & Meng, X. (2017). Understanding the high-electrocatalytic performance of two-dimensional MoS$_2$ nanosheets and their composite materials. *J. Mater. Chem. A* 5, pp. 24540–24563.

Kim, Y. H., Kim, S. J., Kim, Y. J., Shim, Y. S., Kim, S. Y., Hong, B. H. & Jang, H. W. (2015). Self-activated transparent all-graphene gas sensor with endurance to humidity and mechanical bending. *ACS Nano* 9(10), pp. 10453–10460.

Li, J., Han, J., Li, H., Fan, X. & Huang, K. (2020). Large-area, flexible broadband photodetector based on WS$_2$ nanosheets films. *Mater. Sci. Semicond. Process.* 107, p. 104804.

Li, X., Tao, L., Chen, Z., Fang, H., Li, X., Wang, X., Xu, J.-B. & Zhu, H. (2017). Graphene and related two-dimensional materials: Structure-property relationships for electronics and optoelectronics. *Appl. Phys. Rev.* 4, pp. 021306–021337.

Lin, Z., Liu, Y., Yao, Y., Hildreth, O., Li. Z., Moon, K. & Wong, C.-P. (2011). Superior capacitance of functionalized graphene. *J. Phys. Chem. C* 115, pp. 7120–7125.

Liu, B. (2022). Transition metal dichalcogenides for high–Performance aqueous zinc ion batteries. *Batteries* 8(7), p. 62.

Liu, X., Huang, K., Zhao, M., Li, F. & Liu, H. (2020). A modified wrinkle-free MoS$_2$ film transfer method for large area high mobility field effect transistor. *Nanotechnol.* 31, p. 055707.

Liu, Z. & Zhou, X. (2014). Graphene: Energy storage and conversion applications. in: Electrochemical energy storage and conversion. (Taylor & Francis).

Long, M., Wang, P., Fang, H. & Hu, W. (2019). Progress, challenges, and opportunities for 2D material based photodetectors. *Adv. Funct. Mater.* 29, pp. 1803807–1803834.

Luo, B., Liu, G. & Wang, L. (2016). Recent advances in 2D materials for photocatalysis. *Nanoscale* 8, pp. 6904–6920.

Majee, B. P., Bhawna, S. A., Prakash, R. & Mishra, A. K. (2020a). Large area vertically oriented few-layer MoS$_2$ for efficient thermal conduction and optoelectronic applications. *J. Phys. Chem. Lett.* 11, pp. 1268–1275.

Majee, B. P., Mishra, S., Pandey, R. K., Prakash, R. & Mishra, A. K. (2019). Multifunctional few-layer MoS$_2$ for photodetection and surface-enhanced raman spectroscopy application with ultrasensitive and repeatable detectability. *J. Phys. Chem. C* 123, pp. 18071–18078.

Majee, B. P., Srivastava, V. & Mishra, A. K. (2020b). Surface-enhanced Raman scattering detection based on an interconnected network of vertically oriented semiconducting few-layer MoS$_2$ nanosheets. *ACS Appl. Nano Mater.* 3, pp. 4851–4858.

Mas-Ballesté, R., Gómez-Navarro, C., Gómez-Herrero, J. & Zamora, F. (2011). 2D materials: To graphene and beyond. *Nanoscale* 3, pp. 20–30.

Meng, J., Guo, H., Niu, C., Zhao, Y., Xu, L., Li, Q. & Mai, L. (2017). Advances in structure and property optimizations of battery electrode materials. *Joule* 1(3), pp. 522–547.

Meyer, J. C., Geim, A. K., Katsnelson, M. I., Novoselov, K. S., Booth, T. J. & Roth, S. (2007). The structure of suspended graphene sheets. *Nature* 446, pp. 60–63.

Novoselov, K. S., Geim, A. K., Morozov, S. V., Jiang, D. E., Zhang, Y., Dubonos, S. V., Grigorieva, I. V. & Firsov, A. A. (2004). Electric field effect in atomically thin carbon films. *Science* 306(5696), pp. 666–669.

Popa-Simil, L. (2008). Nanotube potential future in nuclear power. *MRS Proc.* 1081, pp. 1015–1081.

Selamneni, V., Nerurkar, N. & Sahatiya, P. (2020). Large area deposition of $MoSe_2$ on paper as a flexible near-infrared photodetector. *IEEE Sens. Lett.* 4(5), pp. 1–4.

Singh, S., Deb, J., Sarkar, U. & Sharma, S. (2020). $MoSe_2$ crystalline nanosheets for room-temperature ammonia sensing. *ACS Appl. Nano Mater.* 3(9), pp. 9375–9384.

Singh, A., Gupta, J. D., Jangra, P. & Mishra, A. K. (2023a). Layered chalcogenides: Evolution from bulk to nano-dimension for renewable energy perspectives. in: Nanomaterials. (Springer) pp. 177–204.

Singh, A., Majee, B. P., Gupta, J. D. & Mishra, A. K. (2023b). Layer dependence of thermally induced quantum confinement and higher order phonon scattering for thermal transport in CVD-Grown triangular MoS_2. *J. Phys. Chem. C* 127(7), pp. 3787–3799.

Sisto, T. J., Zakharov, L. N., White, B. M. & Jasti, R. (2016). Towards pi-extended cycloparaphenylenes as seeds for CNT growth: Investigating strain relieving ring-openings and rearrangements. *Chem. Sci.* 7, pp. 3681–3688.

Song, Y., Li, X., Mackin, C., Zhang, X., Fang, W., Palacios, T. & Kong, J. (2015). Role of interfacial oxide in high-efficiency graphene–silicon Schottky barrier solar cells. *Nano Lett.* 15(3), pp. 2104–2110.

Tan, X., Wu, J., Zhang, K., Peng, X., Sun, L. & Zhong, J. (2013). Nanoindentation models and Young's modulus of monolayer graphene: A molecular dynamics study. *Appl. Phys. Lett.* 102, p. 071908.

Taylor, P., Das, S., Kim, M., Lee, J., Choi, W., Das, S., Kim, M., Lee, J. & Choi, W. (2014). Critical reviews in solid state and materials sciences synthesis, properties, and applications of 2-D materials. in: A comprehensive review. pp. 37–41.

Velraj, S. & Zhu, J. H. (2015). Cycle life limit of carbon-based electrodes for rechargeable metal–air battery application. *J. Electroanal. Chem.* 736, pp. 76–82.

Wang, J., Dong, S., Guo, T., Jin, J. & Sun, J. (2017). pH-dictated synthesis of novel flower-like MoS_2 with augmented natural sunlight photocatalytic activity. *Mater. Lett.* 191, pp. 22–25.

Wang, Q. H., Kalantar-zadeh, K., Kis, A., Coleman, J. N. & Strano, M. S. (2012). Electronic and optoelectronics of two dimensional transition metal dichalcogenides. *Nat. Nanotechnol.* 7, pp. 699–712.

Wigley, T. M. L. & Raper, S. C. B. (2001). Interpretation of high projections for global-mean warming. *Science* 293, pp. 451–454.

Wong, H.-S. P. & Akinwande, D. (2010). Overview of carbon nanotubes, in: *Carbon nanotube and graphene device physics.* (Cambridge University Press, Cambridge) pp. 1–18.

Wu, C. R., Chang, X. R., Chu, T. W., Chen, H. A., Wu, C. H. & Lin, S. Y. (2016). Establishment of 2D crystal heterostructures by sulfurization of sequential transition metal depositions: Preparation, characterization, and selective growth. *Nano Lett.* 16(11), pp. 7093–7097.

Xu, H., Xin, L., Liu, L., Pang, D., Jiao, Y., Cong, R. & Yu, W. (2019). Large area MoS_2/Si heterojunction-based solar cell through sol-gel method. *Mater. Lett.* 238, pp. 13–16.

Yadav, M. K., Panwar, N., Singh, S. & Kumar, P. (2020). Preheated self-aligned graphene oxide for enhanced room temperature hydrogen storage. *Int. J. Hydrog. Energy* 45(38), pp. 19561–19566.

Yang, X., Cheng, C., Wang, Y., Qiu, L. & Li, D. (2013). Liquid-mediated dense integration of graphene materials for compact capacitive energy storage. *Science* 341, pp. 534–537.

Yun, Y. S., Kim, D., Tak, Y. & Jin, H. J. (2011). Porous graphene/carbon nanotube composite cathode for proton exchange membrane fuel cell. *Synth. Met.* 161(21–22), pp. 2460–2465.

Zheng, Z., Zhang, T., Yao, J., Zhang, Y., Xu, J. & Yang, G. (2016). Flexible, transparent and ultra-broadband photodetector based on large-area WSe_2 film for wearable devices. *Nanotechnol.* 27(22), pp. 225501.

Zhou, Z., Bouwman, W. G., Schut, H., van Staveren, T. O., Heijna, M. C. R. & Pappas, C. (2017). Influence of neutron irradiation on the microstructure of nuclear graphite: An X-ray diffraction study. *J. Nucl. Mater.* 487, pp. 323–330.

https://doi.org/10.1142/9789811283406_0002

Chapter 2

Synthesis, Characterization, and Properties of Carbon Nanostructures

Shanu Mishra, Ankita Singh, and Ashish Kumar Mishra

School of Materials Science and Technology
Indian Institute of Technology (Banaras Hindu University),
Varanasi 221005
Email: akmishra.mst@iitbhu.ac.in

Abstract

Carbon nanomaterials, especially graphene and carbon nanotubes (CNTs), are among the most widely studied materials owing to their unique properties spanning from high specific surface area, to excellent electrical and thermal conductivities, flexibility, and optical properties [Smith & Rodrigues, 2015; Yu *et al.*, 2013]. Carbon nanomaterials show significant promises as electrode materials, conductive agents, etc., in energy generation and storage. This chapter focusses on synthesis, characterization, and key properties of carbon nanomaterials such as graphene and CNTs. We have briefly discussed different methods of synthesis of graphene and CNTs and the properties resulting from these methods.

2.1 Introduction

Carbon is a unique and vital material on Earth. The functions of carbon materials are essential due to their outstanding chemical stability, excellent mechanical strength, high electron mobility, and good electrical and thermal conductivities. Also, the hybridization state, large surface area, and highly tunable properties of carbon materials make them suitable for many catalytic processes and energy storage applications. Therefore, carbon materials are very trendy in the fields of nanotechnology, materials science, biotechnology, and engineering sciences. This chapter intends to discuss some common carbon materials (graphene and CNTs) which have effective and efficient properties for environmental applications and energy storage. Their preparation methods along with the characterization techniques and some of the key properties are also discussed briefly.

2.2 Graphene

2.2.1 *Synthesis of Graphene*

Owing to its excellent mechanical, electrical, and chemical properties, graphene has been broadly researched and has gained intense attention in energy storage systems, electronics, water purification, biosensors, biomedical applications, plasmonics, nanoantennas, radio wave absorption, sound transducers, and so on. Several techniques for graphene production have been developed so far. Its synthesis can be carried out by two main approaches: top-down and bottom-up. The top-down approach breaks the larger carbon precursors into graphene by cutting off the layers of graphite by exfoliation (mechanical and chemical), chemical synthesis, arc discharge, and unzipping of carbon nanotubes (CNTs). Usually, this method produces high-quality graphene and has high scalability with the demerit of low yield. To obtain large-scale and defect-free graphene, bottom-up approaches are more favored, where graphene is synthesized from atomic-sized carbon precursors (other than graphite) by using methods such as chemical vapor deposition (CVD) (thermal and plasma enhanced), epitaxial growth on SiC, and pyrolysis. This approach has a demerit of

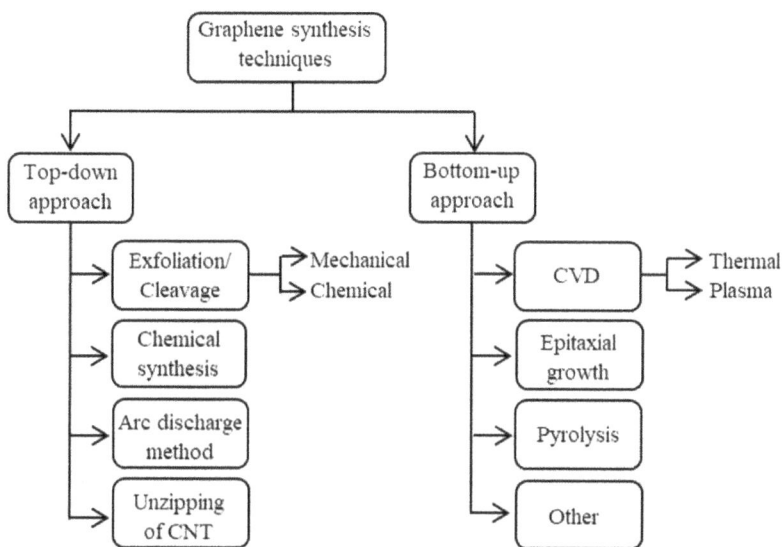

Fig 2.1   Schematic of different synthesis methods of graphene.

high growth cost and a sophisticated operational setup. Other methods include laser irradiation, electrochemical exfoliation, microwave synthesis, conversion of nanodiamonds, etc. An outline of the graphene synthesis technique is shown in Fig 2.1. This chapter focuses on the most commonly used methods.

2.2.1.1 *Exfoliation (or Cleavage) Technique*

Graphite is formed by stacked layers of single-atom-thick graphene bonded by weak van der Waal forces with interlayer separation of 3.3 Å and interbond energy of 2 eV/nm². Exfoliation is the reverse of stacking, that is, slicing down the graphite layers by either applying mechanical or chemical energy to break the weak interlayer bonds. Based on the type of energy applied, exfoliation is categorized into two types: (a) mechanical and (b) chemical. Former was the first recognized technique discovered by Novoselov *et al.* [2004]. An external force of ~300 nNμm⁻² is required to cleave a single layer from graphite [Zhang *et al.*, 2005]. In this method, stress (either longitudinal or transverse) is applied to the graphitic material with the help of a simple scotch tape or an AFM tip, taking out layer by layer. It has also been carried out using an electric field and ultrasonication. Lu *et al.* [1999] first demonstrated the mechanical exfoliation method of plasma-etched pillared HOPG (highly oriented pyrolytic graphite) to produce multilayer graphite (200 nm thick with 500–600 layers) using an AFM tip (Fig 2.2(a)). Novoselov *et al.* [2004] exfoliated single-layer graphene (thickness < 10 nm) from HOPG using scotch tape. They compressed the dry etched (by oxygen plasma) graphite mesa against a wet photoresist layer (1 mm thick) over a glass substrate and baked it to obtain the attachment of HOPG mesa on the photoresist. A graphite flake dispersed in acetone was then deposited onto SiO_2/Si substrate resulting in an atomically thin graphene sheet. Mechanical exfoliation produces high-quality graphene (with lesser defects) but is limited because of the uneven thickness and low yield of the product. The drawback of low production can be overcome using the chemical exfoliation method. In this method, different chemical species are intercalated between the graphite layers, forming graphite intercalation compounds (GICs). It is then reduced to graphene by dispersing in a liquid medium followed by sonication. Intercalation increases the interlayer spacing by reducing the strength of weak van der Waal forces. Alkali metals upon reacting with graphite can easily form GICs because of the difference in their ionization potential and also has the advantage of smaller atomic radii (compared to interlayer spacing of graphite), as shown in Fig 2.2(b). Viculis *et al.* [2005] used this methodology to produce graphene nanoplatelets (GNPs). Potassium was made to react with graphite at 200°C in the presence of an inert atmosphere to form a potassium graphite (KC_8) intercalated compound, followed by its dispersion in aqueous ethanol producing potassium ethoxide ($KOCH_2CH_3$) and H_2 gas that assist the detachment of different layers of graphite, as shown in the following equation:

$$KC_8 + CH_3CH_2OH \rightarrow 8C + KOCH_2CH_3 + \frac{1}{2}H_2 \qquad (2.1)$$

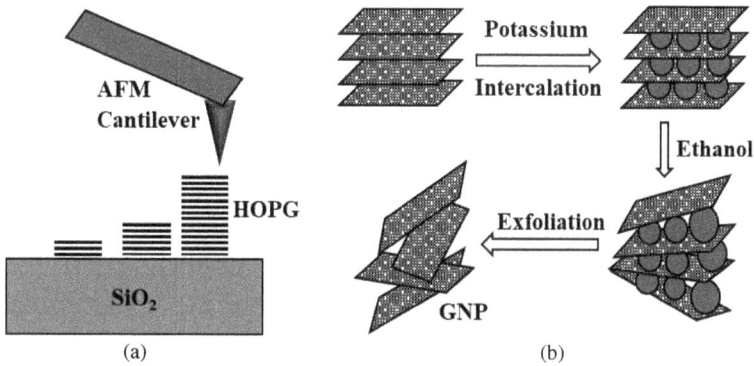

Fig 2.2 Schematic illustrating (a) mechanical exfoliation method and (b) chemical exfoliation process displaying the intercalation of potassium between layers. Fig 2.2(a) adapted from Zhang *et al.* [2005] and Fig 2.2(b) adapted from Viculis *et al.* [2005].

Later, other researchers also employed ferric chloride, nitromethane, and supercritical carbon dioxide as intercalants. The chemical exfoliation method is scalable and produces graphene in huge amounts at low temperatures. But the major demerit of this method is the chemical functionalization of the graphene sheets which disrupts its electronic structure. Although the reduction process eliminates the functional groups, it leaves defects that continue to affect the electronic properties. Hernandez *et al.* [2008] proposed a nonchemical solution-phase method for the exfoliation of graphite using an organic solvent N-methyl-2-pyrrolidone. They synthesized high-quality and unoxidized single-layer graphene with 1 wt% yield, which they further enhanced up to 12 wt% of the initial graphite mass with sediment recycling. The energy needed to peel off the graphene is equalized by the solvent–graphene interaction for the solvents whose surface energies are comparable with graphene. This solution-phase step is versatile, scalable, and can be utilized to form graphene on various substrates. Moreover, graphene-based composites or films can also be produced using this method, which is a crucial requirement for applications in thin-film transistors, conductive transparent electrodes, etc.

2.2.1.2 *Chemical Synthesis*

This is one of the most promising, scalable, low-cost, and broadly studied synthesis methods. It deals with graphene synthesis via a chemical route and comprises the following steps: oxidation of graphite forming graphite oxide (GO) → dispersing by sonication → reducing it to obtain graphene. For GO synthesis, three different methods can be opted: Brodie [1860]; Staudenmaier [1898]; and Hummers and Offeman (1958) shown schematically in Fig 2.3. The oxidation of graphite is carried out in the presence of highly concentrated acids and oxidants, such as sulfuric acid (H_2SO_4), nitric acid (HNO_3), and potassium permanganate ($KMnO_4$). Brodie [1860] first accidentally found an approach to oxidize graphite while investigating the structure of graphite through its reactivity. Mixing potassium chlorate ($KClO_3$) with slurry of graphite in fuming HNO_3 resulted in an overall increment in the mass of graphite flakes with carbon, hydrogen, and oxygen content. Successive oxidation increased the oxygen content with a composition of 1.85% hydrogen, 37.11% oxygen, and 61.04% carbon. It was then dispersed in water or alkaline solutions (not in acid) and heated to 220°C leading to an increase in the carbon content of 80.13%. However, this method was time-consuming and involved several steps.

Staudenmaier [1898] modified the efforts of Brodie by adding $KClO_3$ in multiple parts during the reaction and intensifying the acidity by adding concentrated H_2SO_4 to the process. This process produced highly oxidized GO in one step. A well-known method applied today was readily prepared by William *et al.* [1958] (popularly known as Hummer's method). In this process, GO was synthesized

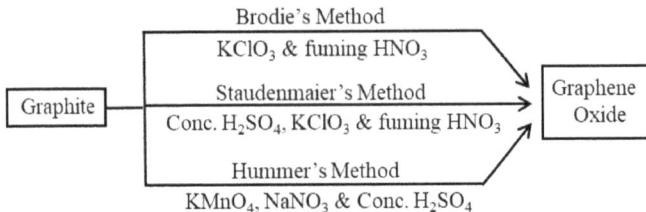

Fig 2.3   Schematic for graphene oxide synthesis from graphite.

by reacting graphite slurry with a mixture of sodium nitrite ($NaNO_3$), concentrated H_2SO_4, and $KMnO_4$. This method is safe and more reliable than the other two due to the excellent oxidation of graphite, though some poisonous gasses such as NO_2 and N_2O_4 are released. To overcome this dilemma, Marcano *et al.* [2010] prepared a modified Hummer's method in which they replaced HNO_3 with H_3PO_4. This efficiently improved the oxidation process and reduced the defects in the basal plane of oxidized graphite. When graphite is turned into GO, interlayer spacing increases depending on the oxidation time. It changes from 3.34 Å in the case of pristine graphite to 5.62 Å after 1 h of oxidation, and further to 7.35 Å after 24 h of oxidation. Finally, GO is reduced to graphene (rGO) by treating with hydrazine hydrate or using reducing agents such as sodium borohydride, hydroxylamine, hydroquinone, and ascorbic acid. [Chua & Pumera, 2014]. Other reported methods include thermal, electrochemical, photochemical, hydrothermal, and microwave-assisted reduction.

2.2.1.3 *Arc Discharge Method*

It is one of the highly adaptable methods for fabricating graphene. The reaction setup consists of a water-cooled, stainless steel cylindrical vacuum chamber and two electrodes, an anode (carbon precursor) and a cathode (graphite rod). Electrodes are connected to an external DC power source. Favorable conditions for producing graphene are high voltage (>50 V), high current (100–150 A), and high hydrogen pressure (>200 Torr) [Rao *et al.*, 2010]. The hydrogen environment in the discharge process ceases the dangling carbon bonds with hydrogen to avoid the synthesis of closed structures (prevent rolling of sheets into nanotubes). The electrodes are kept at a fixed distance of about 1–2 mm and are submerged in either a gas or liquid atmosphere for arc evaporation. When the electrodes are mechanically brought closer to each other, discharge occurs, that is, the medium is dissociated to generate high-temperature plasma, enough to sublime the anode material. When the discharge is over, soot formed on the inner wall of the chamber is collected. It contains only graphene flakes, while the bottom of the chamber comprises other graphitic particles such as MWCNTs, carbon onions, and multilayer graphene. Wang *et al.* [2010] prepared a low-cost and scalable graphene by employing air medium instead of H_2/He medium (as inert gases are generally expensive). However, Wang's method is dependent on air pressure, that is, a high pressure ensures the growth of graphene nanosheets and a low pressure facilitates the formation of other carbon nanostructures. To further decrease the cost, Li *et al.* [2013] synthesized graphene from petroleum asphalt (carbon-rich, inexpensive raw material) using pulsed arc discharge in water. The oil-derived asphalt is the remaining product after the distillation of petroleum and is readily available in nature. The arc discharge method is remarkably used to dope graphene with boron and nitrogen. Subrahmanyam *et al.* [2009] obtained pure B-doped and N-doped graphene in the environment of H_2 mixed with diborane and pyridine, respectively.

2.2.1.4   *Unzipping of CNT*

It is one of the most recent techniques for synthesizing graphene. Narrowing graphene along its width transforms the electronic properties from semimetal to semiconductor [Chen *et al.*, 2007]. In this method, starting material (single- or multiwalled CNT) determines whether the formed graphene is single or multilayer. On unzipping CNT, a strip of graphene possessing straight edges is obtained, known as graphene nanoribbons (GNRs), and the diameter of the nanotube determines the width of the nanoribbon. The unzipping of CNT is done using chemical and plasma-etching methods.

Cano-Márquez *et al.* [2009] reported a new chemical technique for longitudinal unzipping of MWCNTs by intercalating with lithium and ammonia followed by peeling off with intense acid and heat treatment. They dispersed CVD-synthesized MWCNTs in dry tetrahydrofuran (THF) and then liquid NH_3 was added. The temperature of the reaction flask was retained at –77°C using an acetone-dry ice bath. Next, Li was added in a proportion of 10:1 (Li:C) and was left for a few hours to allow the intercalation of MWCNT. Later, HCl was slowly added to this solution for exfoliation (highly exothermic) to occur and finally obtained using a microfiltration membrane. This produced ~60 % fully exfoliated and (0–5%) partially exfoliated or damaged MWCNTs. Unexfoliated or partially exfoliated CNTs were further exfoliated by thermal treatments. At the same time, Kosynkin *et al.* [2009] also described a solution-based oxidative step of CNTs' side wall followed by its unraveling. First, they oxidized nanoribbons by suspending MWCNTs in concentrated H_2SO_4 for a duration of 1–12 h. Then it was made to react with $KMnO_4$ for 1 h each at 22°C and 55–70°C, respectively. After the entire $KMnO_4$ was consumed, the reaction mixture was quenched with ice impregnated with H_2O_2. Then the solution was seeped through a polytetrafluoroethylene (PTFE) membrane followed by subsequent washing. It was further reduced using concentrated ammonium hydroxide (NH_4OH) and hydrazine monohydrate ($N_2H_4 \cdot H_2O$) to regain its electrical feature. In the beginning, the diameter of MWCNTs was 40–80 nm and after opening the thickness of GNR, it enlarged to 100 nm. Tanaka *et al.* [2015] fabricated single-layer GNRs by using double-walled CNTs (DWCNTs). They observed that DWCNTs are better starting materials than MWCNTs and that unzipping of DWCNTs produces bilayer graphene, which can further be separated into two single-layer graphenes via sonication. Fig 2.4 shows the procedure for unzipping DWCNTs to form single-layer GNRs. First, the defects were induced into DWCNTs by annealing at 500°C. Then they were dispersed in

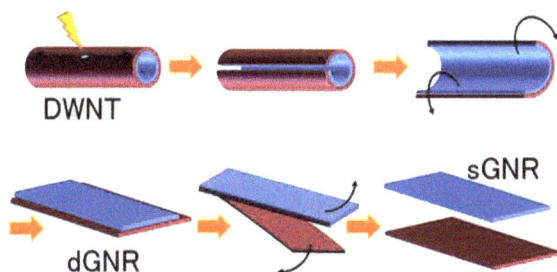

Fig 2.4   Schematic showing the procedure of getting graphene nanoribbon from DWCNT. Reprinted with permission from Lee *et al.* [2017]. Copyright (2023) Royal Society of Chemistry.

organic solutions, followed by sonication which leads to the formation of single-layer GNR.

2.2.1.5 *Chemical Vapor Deposition*

Among all methods, CVD yields high-quality products with low defects, uniform and continuous monolayer, as well as few-layer graphene. CVD works by transporting the gas precursors (using carrier gas) in the reaction zone, where the heated substrate is placed and a chemical reaction occurs forming a thin-film deposit on the substrate. In the case of solid/liquid precursor, it is first vaporized followed by deposition via condensation onto the substrate. In simple words, the synthesis of graphene by CVD method mainly involves the following two steps:

(i) pyrolysis of gas precursor to form carbon atoms,
(ii) formation of carbon structure, i.e., graphene onto the heated substrate.

Graphene film can be synthesized via this method on metal substrates such as Ni, Fe, Cu, Pt, Ir, and Ge by employing carbon sources such as CH_4, C_2H_2, and C_6H_{14}. In situations where high-temperature deposition is not attainable, plasma-assisted deposition is employed. Thus, depending on the growth condition, available precursor, desired quality, and thickness of graphene film required, there are mainly two common types of CVDs available: thermal CVD and plasma-enhanced CVD (PECVD).

a) Thermal CVD: In this process, thermally decomposed precursors are exposed to a substrate at high temperature leading to thin film deposition on the substrate upon cooling. Preset conditions determine the number of layers formed. Many factors affect the growth process such as the choice of substrate, gas flow rate, cooling rate, reaction temperature, reaction time, and so on. The optimization of various growth parameters is important. Fig 2.5 shows the schematic CVD setup for preparing graphene. Graphene grown on

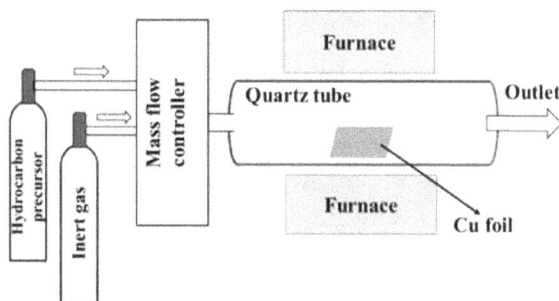

Fig 2.5   Schematic setup for growing graphene on Cu foil via CVD.

Cu substrate has attracted greater interest because of its low cost, lower solubility of carbon, and ease of transfer onto the desired substrate. Li *et al.* [2009] were the first to develop uniform and large-area graphene synthesis process on Cu foil at 1000°C by CVD using methane and hydrogen. Film growth is described by surface catalyzed process and <5% of the area is 2–3 layers. Poor carbon saturation due to graphene coverage on the surface also limits the precipitation process. Likewise, Lee *et al.* [2015] reported an improved method for growing graphene on Cu through CVD. Short growth time, increased growth pressure, and low flow rate of the H_2/CH_4 gas mixture are maintained to enhance the film quality. Coalescence during the pre-annealing step reduces the surface roughness and steps of Cu grains to a great extent making it relatively smoother. The variation in the thermal expansion coefficients of Cu (16.3×10^{-6} K^{-1} at 300 K and 20×10^{-6} K^{-1} at 800 K) and graphene (-7.9×10^{-6} to 4.35×10^{-6} K^{-1}) produces ~0.3–0.6 % thermal strain in graphene, which is less than the graphene's intrinsic strain. As a result, some areas of graphene are lifted producing folding lines.

The reaction occurs in two phases: first, the breakdown of hydrocarbon to form active carbon species. Next, these active carbon species as well as unreacted hydrocarbon adsorb on the substrate (Cu) and undergo further reactions to form graphene. For practical applications, a reliable method with high efficiency and low cost is necessary for the damage-free transfer of graphene film onto the desired substrate. Till now, the common method used for transferring

CVD-grown graphene from Cu to the required substrate is the polymer-assisted transfer process. In this, a polymer (polymethyl methacrylate (PMMA), polydimethylsiloxane (PDMS), or polystyrene (PS)) layer is used as a temporary supportive framework on the graphene film to avoid its folding or damage during etching. It has the merit of easy handling and processing.

b)   Plasma-Enhanced CVD: In this method, precursor gases undergo a chemical reaction inside a vacuum chamber by employing plasma, thus producing a thin film onto the substrate. The source for generating plasma can be microwave, radio frequency (RF), or inductive coupling. This method is more beneficial than other CVD methods for large-scale industrial purposes because of lower growth temperature, shorter deposition time, and catalyst-free graphene synthesis. Wang *et al.* [2004] have shown the growth of graphene by RF-PECVD on various substrates such as Si, SiO_2, Al_2O_3, Mo, Zr, Ti, Hf, Nb, W, Ta, Cu, and 304 stainless steel under the same deposition condition and observed the same morphology of nanosheets on all the substrates irrespective of its variation. The chamber pressure was kept at 12 Pa, 5–100 % CH_4 in H_2 with 10 sccm flow rate, temperature ranging from 600°C to 900°C with growth time of 5–40 min, and 900-W RF power. High-quality sheets of graphene (1–3 layers) were synthesized by microwave plasma CVD on stainless steel substrates at 500°C in a CH_4/H_2 gas mixture environment (in 1:9 ratio with 200 sccm flow rate and total pressure of 30 Torr), and a microwave power of 1200 W during growth [Yuan *et al.*, 2009]. This approach produced graphene with high yield and high purity with no contamination.

2.2.1.6 *Epitaxial Growth on SiC*

This is one of the most appreciated techniques for growing high-quality monolayer graphene. This method allows the growth of single-crystalline film over a single-crystal SiC substrate (nonconducting). When the epitaxial film and the substrate are the same, then it is referred to as a homoepitaxial layer, else called hetero-epitaxial layer, for example, epitaxial formation of graphene on SiC produces a heteroepitaxial layer. Berger *et al.* [2004] produced graphene (1–3 layers) grown epitaxially on the Si-terminated (0001) face of single-crystal 6H-SiC. The process consists of following steps: (a) surface quality improvement by oxidation or H_2 etching; (b) oxide removal by heating the sample at ~1000°C in ultrahigh vacuum (UHV) using electron bombardment; (c) heating the sample at a temperature ranging from 1250°C to –1450°C for 1–20 min. The thickness of the grown film is mainly determined by the temperature. In 2012, the epitaxial graphene was grown on Ni thin film–coated SiC substrate [Juang *et al.*, 2009]. The process involves coating of 200 nm Ni on single-crystalline 6H–SiC (0001) and 3C–SiC substrates by electron beam evaporation. Then the sample was heated at ~750°C under ~10^{-7} Torr pressure. Graphene fabricated by this method is continuous over the entire Ni-coated region and can be easily transferred onto another substrate. This method is very promising due to high-quality, large-scale, and low-temperature growth.

2.2.1.7 *Pyrolysis*

This method deals with the chemical synthesis of graphene via a bottom-up approach, known as the solvothermal method. All solvothermal reactions are carried out in a Teflon-lined reactor. The synthesis process involves the heating of sodium and ethanol in the molar ratio of 1:1 in a closed container at 220°C for 72 h, forming a solid precursor (sodium ethoxide) [Bhuyan *et al.*, 2016]. Then the resultant solid is rapidly pyrolyzed to detach the graphene sheets, vacuum filtered, and dried using a vacuum oven at 100°C for 24 h. This synthesis technique yields 0.1 g graphene per 1 ml of ethanol, that is, 0.5 g per solvothermal reaction. The dimensions of synthesized graphene were up to 10 μm. Raman spectra of the grown graphene result in a broad D-band with I_G/I_D ~1.16, showing defective graphene. Although this method is cost-effective, scalable, and low-temperature, it produces functionalized graphene with a large number of defects.

2.2.2 Characterization Techniques of Graphene

The synthesized graphene would be characterized by various microscopic and spectroscopic techniques to detect the morphology, structure, thickness, quality, impurity, and defects. Often used characterization techniques consist of optical microscopy, Raman spectroscopy, electron microscopy, and scanning probe microscopy.

2.2.2.1 *Optical Microscopy*

It is an easy, straightforward, and nondestructive characterization technique for large-area samples, primarily used to estimate the shape, size, and number of layers present in the graphene film. The analysis is done by observing different color contrasts between the graphene and the underneath substrate (dielectric) due to interference and diffraction of light. In the past decade, imaging of fluorescent single layers was achieved using a thin dielectric layer between the sample and the substrate. Fabry-Perot interference in the dielectric layer allowed the estimation of the sample thickness [Jung *et al.*, 2007]. Generally, silicon dioxide (SiO_2) and silicon nitride (Si_3N_4) are deposited on Si because of their contrast-enhancing dielectric properties. A similar method was adapted for the recognition of layer numbers in graphene deposited on SiO_2/Si substrate (300 nm thick) and identified by white light illumination. The contrast was enhanced by narrowing the band illumination. According to Fresnel theory, the contrast of graphene film on SiO_2/Si substrate depends on the wavelength of incident light and thickness of the dielectric [Blake *et al.*, 2007]. Fig 2.6(a) shows the optical image of CVD-grown monolayer graphene, electrochemically transferred on SiO_2/Si substrate. The arrow shows the bilayer graphene islands [Lee *et al.*, 2017]. Ni *et al.* [2007] calculated the contrast spectra between the graphene film and the substrate, given by the following equation:

$$C(\lambda) = \frac{R_0(\lambda) - R(\lambda)}{R_0(\lambda)} \qquad (2.2)$$

where $R_0(\lambda)$ and $R(\lambda)$ are the reflection spectrum from the SiO_2/Si substrate and from the graphene film, respectively. Using this equation, the number of graphene layers present on a 300-nm-thick SiO_2/Si substrate can be determined by $C = 0.0046 + 0.0925\ N + 0.00255\ N^2$, where $N\ (\leq 10)$ is the number of graphene layers.

2.2.2.2 *Scanning Electron Microscopy*

SEM is used as an effective tool for the determination of surface morphology of graphene samples. This is done by focusing a highly energetic beam of electrons on the sample. These accelerated electrons have a much shorter wavelength resulting in high resolution in SEM and field emission SEM (FESEM) as compared to optical microscopy. Fig 2.6(b) shows the SEM image of graphene revealing the continuous monolayer growth with some ripples and bilayer islands [Lee *et al.*, 2017]. In the case of few-layer graphene (FLG), SEM image shows better contrast compared to optical image. Van Khai *et al.* [2013] synthesized FLG with some oxygen content by microwave-assisted solvothermal technique. FESEM image shows that the FLG sheets are randomly individual and separate with sizes ranging from 3 to 10 μm. Also, wrinkles on the surface and folding at the edges are clearly visible. Wrinkles may be present due to the remaining oxygen-containing functional groups (e.g., -COOH and-OH) present on the sides of the graphene film. Takahashi *et al.* [2012] employed SEM to examine the in situ graphene formation on a polycrystalline Ni substrate. Graphene with variation in layer numbers was synthesized and distinguished based on contrast, that is, 1–2L graphene film appeared brighter, 3–4L graphene showed an intermediate contrast and thicker graphene layer appeared darker. This contrast change was observed by the distinctness in the work function, that is, for Ni(111) surface it is 5.3 eV, for single-layer graphene-covered surface it is 3.9 eV, and for the graphite surface it is 4.6 eV, and also due to the variation in the number of valence electrons for Ni and graphite. The difference in contrast was enhanced when the sample was brought to room temperature in vacuum.

Air exposure of the sample was done to well distinguish the monolayer graphene from the bare substrate. The oxidized bare Ni surface (increases secondary electron emission from the oxidized areas) seemed to be brighter in the SEM image than the graphene-covered surface which is resistant to oxidation.

2.2.2.3 *Scanning Probe Microscopy*

It characterizes the nanomaterial's topography by scanning the sample with a sharp nanometer probe. Two frequently used SPM modes for characterizing graphene based on their sensing mechanism are atomic force microscopy (AFM) and scanning tunneling microscopy (STM). In AFM, the tip attached to the adjustable cantilever deflects on sensing the surface of the sample to form 2D and 3D images. Scanning in the AFM mode can be executed in two ways: contact or noncontact tapping mode. In contact mode, the probe just touches the surface, whereas in the noncontact mode it hovers above the surface of the sample, and a small attractive force acts between the tip and the sample constructing the topographic images.

AFM is a useful method to verify the thickness of graphene. For monolayer graphene, the thickness is around 0.34 nm. Pristine graphene and GO can be differentiated on the basis of their thicknesses determined with AFM imaging. Paredes *et al.* [2009] employed phase imaging in the attractive region of tapping-mode AFM to distinguish the chemically reduced and unreduced GO nanosheets. They obtained a thickness of ~1.0 and ~0.6 nm for the unreduced GO and chemically reduced, respectively. This significant variation is due to hydrophilicity appearing from a distinct oxygen functional group on the unreduced GO. Fig 2.6(c) displays the AFM image and height profile of monolayer graphene on SiO_2/Si substrate with little cracks [Lee *et al.*, 2017]. Apart from characterization, AFM is also applied in other useful works such as nanolithography, patterning, and cutting of graphene, etc. The layer number of graphene, defects, rotational disorder between graphene layers, and substrate graphene mismatch can be estimated using STM [Wong *et al.*, 2012]. Characterizing graphene using STM produces high-contrast atomic resolution images showing hexagonal close-packed lattice structures.

2.2.2.4 *Transmission Electron Microscopy (TEM)*

This is a very impressive tool for the structural characterization of graphene. It mainly gives a focused and magnified image by allowing the electron beam to pass through the sample. A thin film sample is preferred for its analysis and graphene being ultrathin can be directly detected by TEM. It can accurately estimate the thickness of graphene. High-resolution TEM (HRTEM) can straight away characterize graphene at the atomic level (e.g., point defects, Stone–Wales rotation, vacancy, dislocations) up to 1Å resolutions under a low voltage electron beam, as high operational voltage induces defects in the graphene. TEM examination gives a correct approach for the analysis of layer number in graphene films grown according to the synthesis conditions. The selected area electron diffraction (SAED) pattern discloses the typical hexagonal crystalline nature of graphene [Tu *et al.*, 2014]. SAED pattern of bilayer graphene consists of 12 spots in each ring due to the contribution from 2 layers rather than just 6 as in the case of monolayer graphene. Fig 2.6(d–f) shows the high-magnification TEM images of different thicknesses of graphene (bilayer, trilayer, and four-layer) synthesized via CVD method [Lee *et al.*, 2017]. In this report, the number of layers was obtained by observing cross-sectional TEM images of graphene samples. Similarly, Meyer *et al.* [2008] illustrated direct imaging that resolved every individual carbon atom in a suspended monolayer graphene lattice using HRTEM. They acquired 1Å resolution in HRTEM at an 80 kV of acceleration voltage by using aberration correction with a monochromator. They also studied the imperfections (e.g., defects, vacancies, edges, and adsorbates) in single-layer graphene. The grain boundaries along with their defect structure can also be visualized showing the joining of two grains using HRTEM. Apart from visualizing atoms, defects, and atomic arrangements in graphene by HRTEM, it has a limitation on sample preparation which requires expertise. Being expensive, it needs precise control over the instrument to obtain a clear atomic resolution picture of graphene.

2.2.2.5 *Raman Spectroscopy*

It is a powerful nondestructive characterization tool used for the qualitative and quantitative determination of carbonaceous materials. The difference in frequency between the incident light and scattered light after interaction with the sample is due to the rotation and vibration of the molecules in the sample, thus determining the molecular structure of the material. It is used for the analysis of quality of film, layer number, defects, and the induced strain present in graphene. Carbon allotropes possess unique Raman fingerprints with characteristic D-band, G-band, and 2D-band at 1350 cm^{-1}, 1580 cm^{-1}, and 2700 cm^{-1}, respectively [Gupta *et al.*, 2006; Yoon *et al.*, 2009]. Fig 2.6(g) shows the comparative Raman spectra of graphene and graphite. The G-band originates from the in-plane vibrations of sp^2-bonded carbon atoms, that is, the tangential stretching (E_{2g}) mode of HOPG. It mainly deals with the degree of crystallization and the symmetry of graphene. As the layer number increases, so does the intensity of the G-band, being the lowest for monolayer graphene. On the other hand, the D-band originates from the out-of-plane vibrations (breathing mode), that is, due to the disorderness present in the sp^2-hybridized carbon atoms and has main attributions from lattice disorders, structural defects, wrinkles, edge effect, etc. The absence of the D-band in the Raman spectra shows the formation of a defect-free film. Graphene synthesized using the chemical reduction method shows a large no. of defects compared to that synthesized by any other technique. The 2D-band (also labeled as G' band in some studies) appears at twice the frequency of the D-band and arises from two-phonon double-resonance Raman scattering (known as two-phonon scattering) [Thomsen & Reich, 2000]. Both the D- and 2D-bands are second-order Raman scattering phenomena with the difference that D-band arises due to an elastic scattering and an inelastic scattering, whereas the 2D-band is due to two inelastic scatterings. Positions and shapes of the G- and 2D-bands change as the layer

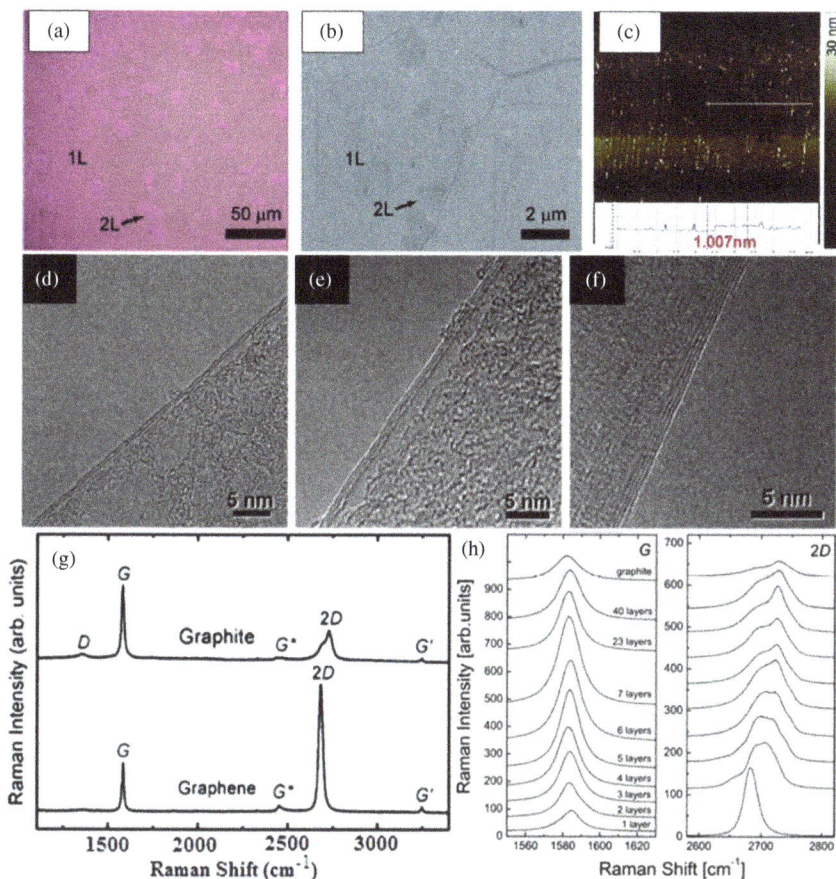

Fig 2.6 (a) Optical image, (b) SEM, (c) AFM of single-layer graphene, electrochemically transferred on SiO_2/Si substrate, (d–f) HR-TEM images of graphene edges showing two, three, and four layers, respectively, (g) Comparative Raman spectra of graphene and graphite and (h) Evolution of G- and 2D-bands in the Raman spectra with increasing layer number of graphene. Reprinted with permission from Lee *et al.* [2017]. Copyright (2023) Royal Society of Chemistry.

number increases, as shown in Fig 2.6(h). The 2D-band in 1L graphene has a single component, whereas 2L, 3L, and 4L graphene involves two or more components. The ratio I_D/I_G assesses the extent of disorder present in graphene and the I_G/I_{2D} ratio signifies the quality of as-grown graphene by giving the layer number. I_G/I_{2D} for monolayer graphene is ~0.24 and increases further with layer number [Das *et al.*, 2008]. Hence, a small value of I_G/I_{2D} and a slight hump of the D peak imply the high quality of synthesized graphene. Thus, Raman spectroscopy can clearly differentiate 1L, 2L, and multilayer graphenes.

2.2.3 *Properties of Graphene*

2.2.3.1 *Electronic Properties*

The electronic property of graphene is closely related to its π bond. The charge carriers in graphene can be described from the Dirac's (relativistic) equation and are massless near the Dirac point. Due to the honeycomb lattice consisting of 2 atoms per unit cells, the cone-like valence band (VB) and conduction band (CB) intersect at the K and K′ points of the Brillouin zone to form the Fermi level. At this point, the energy band structure of graphene shows linear dispersion relation. Graphene is considered semimetallic due to its zero bandgap and zero density of state of electrons at the Fermi level. It consists of bipolar conduction characteristics with a high concentration of 10^{13} cm^{-2}. It features high carrier mobility of $\approx 500,000$ cm^2 V^{-1} s^{-1}, which can be modulated by inducing chemical doping on the surface [Jo *et al.*, 2012]. Doping with *n*-type makes the upward shift of the Fermi level, thus increasing the conductivity and decreasing the work function. While doping with *p*-type also increases the conductivity but downward shifting of the Fermi level leads to an increase in the work function. Therefore, chemical doping changes the Fermi level and hence changes the electronic property that is useful in various applications.

2.2.3.2 *Mechanical Properties*

The mechanical properties of 2D materials play a crucial role in manufacturing and performance in their different applications. This property mainly involves Young's modulus and fracture strength.

Graphene is the strongest ever-known material (200 times that of steel) with 1 TPa measured value of Young's modulus and 130 GPa of fracture strength. These two high values make the hexagonal structure of graphene very strong and rigid. Owing to these two intrinsic properties, graphene is suitable for different applications such as resonators and pressure sensors. It is also a very light material weighing about 0.77 mg/m². It also has an elastic property that retains its original size after applying strain (high flexibility). Its mechanical properties can be macroscopically explained using continuum elasticity theory. Lee *et al.* [2008] measured the mechanical properties of suspended single-layer graphene by AFM nanoindentation. They found the breaking strength of 42 N m⁻¹ and Young's modulus to be 1.0 ± 0.1 TPa (thickness assumed is 0.335 nm). The elastic modulus of the monolayer graphene sheet obtained by reducing graphene oxide with a hydrogen plasma is estimated to be about 0.25 TPa [Zhu *et al.*, 2010]. This superior property makes graphene a promising candidate for applications in nanoelectromechanical systems or other flexible papers.

2.2.3.3 *Optical Property*

Growing world demands for a flexible transparent electrode for applications in touch panels, flexible displays, solar cells, thin film photovoltaics, etc., and Indium tin oxide (ITO) is employed for this rising demand. However, ITO has a limitation of high cost, which demands an alternative. Thus, graphene has attracted attention as a promising candidate. Graphene being conducting and transparent finds application in many photonic devices. The optical absorption of light is found to increase linearly with the layer number. For monolayer graphene, absorption can be calculated as $A = 1 - T = \pi a = 2.3\%$ of white light, where T is the transmittance given by $T = (1 + 0.5n\alpha)^{-2} \approx 1 - n\alpha \approx 97.7\%$ and $a = 1/37$ is the fine structure constant [Nair *et al.*, 2008]. The next important factor is the sheet resistance (R_s) that relies upon the surface morphology and crystal quality of graphene, which varies with different synthesis methods. Among all, graphene synthesized by CVD has the lowest R_s value which can be further reduced by doping. However, transparency and charge carrier mobility must not be affected during doping. Also, R_s decreases with increasing number of layers.

2.2.3.4 *Thermal Properties*

Strong in-plane carbon bonding and negligible phonon scattering in 2D graphene are accountable for its high thermal response. The highest-ever thermal conductivity is observed in single-layer graphene. Due to its high value, it is proposed in electronic applications for thermal management. Balandin *et al.* [2008] calculated the room temperature thermal conductivity of suspended monolayer graphene to be in the range ~$(4.84 \pm 0.44) \times 10^3$ to $(5.30 \pm 0.48) \times 10^3$ W/mK. Such extreme values make graphene outperform CNTs (3000 W/mK for MWCNT and 3500 W/mK for SWCNT). Its value is controlled by the type of structure (AA or AB) and the number of layers. Thermal conductivity is degraded by the presence of defects, edge scattering, and doping. However, the thermal conductivity of supported graphene varies with support and requires further investigation.

2.2.3.5 *Electrochemical Property*

Reduced graphene oxide (rGO) and/or few-layer graphene structures have unique electrochemical properties that make them promising for various applications including energy storage, electrocatalysis, and sensing. rGO has a high electrical conductivity due to the presence of sp^2-hybridized carbon atoms, large surface area due to its two-dimensional structure, and the presence of defects, such as vacancies and functional groups. These properties make it an excellent material for use as an electrode in energy storage devices such as batteries and supercapacitors. It also shows high electrochemical stability due to its reduced oxygen content and the removal of functional groups during the reduction process. This stability makes rGO a suitable material for use in harsh electrochemical environments. rGO can serve as a support for various electrocatalysts, such as metal nanoparticles or metal oxides, which can be deposited on its surface. The presence of rGO can enhance the electrocatalytic activity of these materials by providing conductive and stable support for catalytic application [Smith *et al.*, 2019].

2.3 Carbon Nanotubes

2.3.1 *Synthesis of CNTs*

CNTs have been vigorously studied since their discovery by Iijima [1991] owing to their structural, electrical, and mechanical properties. The nanotubes are formed when one or more layers of graphene sheet are curled into seamless cylinders forming single-walled CNTs (SWCNTs) or multiwalled CNTs (MWCNTs) [Iijima & Ichihashi, 1993; Tasis *et al.*, 2006]. The SWCNT is a tubular form of single-layer graphene, whereas MWCNTs are made up of many concentric tubes of graphene nested within one another. Typically, the diameters of CNTs are in the range of a few angstroms to tens of nanometers, although those of MWCNTs can exceed 100 nm and a length of up to several micrometers. The substantial quantity of CNTs can be synthesized via several techniques that are displayed in Fig 2.7, each having some pros and cons producing different types of nanotubes. A wide range of approaches have been developed to produce CNTs, for

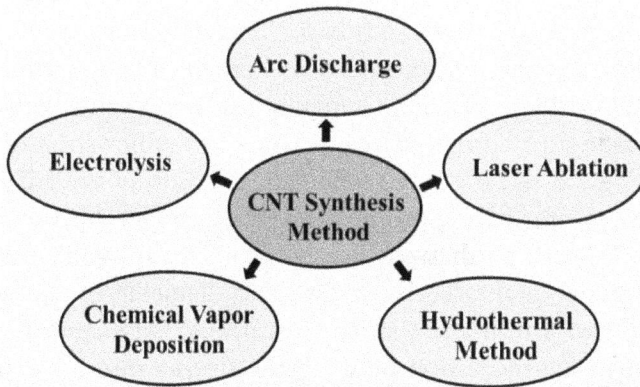

Fig 2.7   Methods used for CNT synthesis.

example, arc-discharge or laser ablation, which are high-temperature preparation techniques resulting in complicated equipment and low efficiency.

Nowadays, low-temperature (<800°C) synthesis techniques, such as CVD, have substituted the high-temperature ones which use catalytic decomposition of hydrocarbons and offer tune orientation, well aligned (length and diameter), high efficiency, and high purity product. Mostly, these methods rely on supporting gases and vacuum, so these volumetric methods are suitable for large-scale synthesis. The growth of CNTs at atmospheric pressure is also reported in the literature [Nozaki & Okazaki, 2008; Nozaki *et al.*, 2011]. Most of the abovementioned techniques produce a small amount of powder CNT with other carbonaceous impurities such as amorphous carbon, nanocrystalline graphite, fullerenes, and metals (Fe, Co, Ni, etc.). These impurities hinder the most desired properties of CNTs, and therefore efficient and simple purification methods are followed after the synthesis of CNT, such as acid treatment [Unrau *et al.*, 2010].

2.3.1.1 *Arc Discharge*

The arc plasma evaporation of pure graphite rod was discovered by Iijima for the synthesis of CNTs [Iijima, 1991]. The method typically uses higher temperatures for CNTs production and provides CNTs with minimum structural defects compared to other synthesis techniques. The principle method is based on establishing a DC electric discharge in an inert environment (helium or argon) between two graphitic electrodes. The high temperature transpiring between the two electrodes during the process permits the sublimation of carbon material. Fig 2.8 illustrates the schematic diagram of the arc evaporation apparatus for CNT production [Kingston & Simard, 2003]. Here, CNTs were synthesized by applying DC arc voltage between two high-purity graphite electrodes. The diameter of the anode is between 6 and 8 mm and that of the cathode is twice of the anode (8–12 mm). The separation between the tip is around 1–2 mm and the electrode chamber is filled with inert gas (helium or argon or

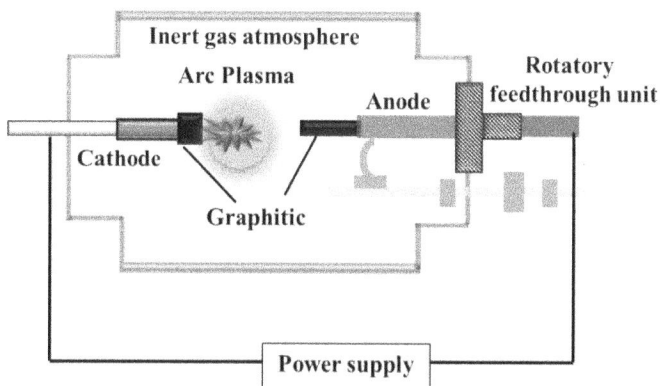

Fig 2.8   Schematic diagram showing the arc discharge setup. Adapted from Kingston & Simard [2003].

hydrogen) at subatmospheric pressure [Zhao *et al.*, 1997; Ajayan *et al.*, 1993]. The systematic water cooling of the cathode is also necessary for obtaining quality nanotubes. The anode's position is controlled from outside the chamber for maintaining a proper gap between the electrodes during arcing. An arc is generated in between the electrodes by passing the direct current ~100–200 A and a bias of ~20–30 V through the camber. This arc current produces plasma of very high temperature ~4000–6000 K that sublimes the carbon precursor present in the anode. While arcing, the positive electrode (anode) is consumed, and the evaporated carbon deposits on the cathode in the form of CNT and other forms of carbon. There are two different ways to synthesize CNTs with the arc discharge method either via using different catalyst precursors or without using catalyst precursors, for example, synthesis of MWCNTs uses no catalyst precursors but the synthesis of SWCNTs utilizes different catalyst precursors. The arc discharge utilizes different compositions of electrodes such as graphite and metal or metal combination (Gd, Co, Ni, Fe or Co-Pt, Fe-Ni, etc.).

The arc-discharge technique has the potential to provide a large concentration of nanotubes. But this method has minute control over the chirality of the synthesized nanotubes which holds major importance on its practical uses.

2.3.1.2 *Laser Ablation*

Laser ablation is a superior method for growing high-quality and high-purity SWCNTs, developed at Rice University by Guo *et al.* [1995]. They used a classic setup of oven laser-vaporization apparatus (Fig 2.9). The working principles and mechanism of laser ablation are similar to arc discharge with the exception that in this case the energy is provided by a high-power laser (YAG-type) targeted on a metal-graphite composite kept in a high temperature (1200°C) furnace in an inert atmosphere. Most of the carbon species produced by laser vaporization were swept out of the furnace by flowing carrier gas (Ar) from high-temperature region and deposited on a conical water-cooled Cu collector. The studies have shown that the tube diameter relies on laser power, that is, a thinner tube diameter is observed with increasing laser power. There are many variables such as chemical configuration of target material, laser properties, and chamber pressure, which influence the properties of nanotubes produced by the laser ablation method. The method provides relatively low metal impurity in synthesized CNTs because the atoms involved tend to evaporate at the end of the tube. The main disadvantages of this

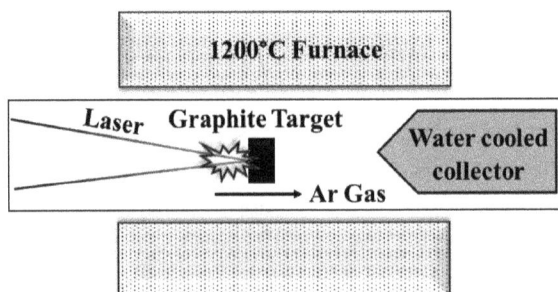

Fig 2.9 Schematic diagram showing the oven laser-vaporization apparatus for nanotube synthesis. Adapted from Journet *et al.* [2012].

technique are the production of branched nanotubes and relative functioning complexity.

2.3.1.3 *Chemical Vapor Deposition*

The arc discharge method produces large quantities of unpurified nanotubes and therefore significant efforts are continuously made for more controllable methods. CVD is a process that offers controllable and selective production of nanotubes. There are different CVD types, such as catalytic CVD (CCVD), oxygen/water-assisted CVD, hot filament CVD, and microwave plasma CVD. CCVD is an economically viable process for obtaining quite pure CNTs, has easy controllability, and can be used for large-scale synthesis. Fig 2.5 shows the schematic diagram of a CVD setup. In this system, nanotubes are synthesized via pyrolytic decomposition of hydrocarbons (benzene, xylene, acetylene, etc.) used as the carbon source at a specific temperature ranging from 500°C to 1200°C either on a substrate (carbon, quartz, silicon) or on metal catalyst particle (Fe, Ni, Co, etc.).

The synthesis process is carried out in a heater/reactor. Here, a small amount of the catalyst is kept in a ceramic or quartz boat which is further placed in a cylindrical quartz tube. In the initial step of nanotube growth, two types of gases are fuelled in the reactor: one is carbonaceous gas (ferrocene, benzene, acetylene, methane vapors, etc.) and the second is inert gas (nitrogen, hydrogen, etc.). The vapors of a carbon-containing gas break at the surface of the catalyst (Fe, Ni, Co, etc.), and the carbon particles become visible at the edge of the catalyst where nanotubes are formed. The growth of the nanostructures takes place in the temperature range of 500–1200°C. The system is then cooled down to room temperature. CCVD is an economically viable and practical method for CNT production with the advantage of easy control of the reaction course.

2.3.1.4 *Hydrothermal Process*

The hydrothermal technique has been utilized for the commercial production of advanced engineering materials such as nanorods, nanowires, nanobelts, nanotubes, and others. This synthesis technique has numerous advantages in contrast to others: environmentally friendly, easy availability of reactive precursors, no initial catalyst requirement, low synthesis temperature (about 150–180°C for the entire process), and absence of carrier gas to run the system. Manafi *et al.* [2008] have synthesized large quantities of CNTs by a hydrothermal method using dichloromethane, metallic lithium, cobalt chloride, and NaOH solution as the starting materials at 150–160°C for 24 h. Krishnamurthy and Namitha [2013] synthesized MWCNT via a hydrothermal process at low temperature of 200°C using ferrocene which acts as both carbon precursor and catalyst. In a typical synthesis, a 1:2 ratio of ferrocene and sulfur was dissolved in a mixed solution of water, ethanol, and NaOH maintained at 200°C for 20 h. The synthesized nanotubes have a range of several hundreds of nm to μm.

2.3.1.5 *Electrolysis*

Electrolysis is a unique method of nanotube synthesis because it transpires in a condensed phase using a graphitic rod at comparatively low synthesis temperatures. This method was proposed by Hsu *et al.* [1995] for CNT production by passing the electric current between carbon electrodes in molten lithium chloride. The fundamental point of this approach is the electrodeposition of alkali metals from their chloride salts onto a graphite cathode which further results in the formation of a CNT.

Dimitrov *et al.* [2011; 2013] performed a constant voltage molten salt electrolysis rather than the constant current electrolysis as used in previous cases to increase the CNT production yield. The graphitic cathode deteriorates during the electrolysis and releases carbon constituents from the surface, thus forming CNTs within the electrolyte. This technique can synthesize MWCNTs with diameters of 10–50 nm and a length of 10–100 μm. Other carbonaceous materials are also produced in this process such as carbon fibers, encapsulated particles, and amorphous carbon. After electrolysis, molten salt is cooled down and the obtained carbon product along with the

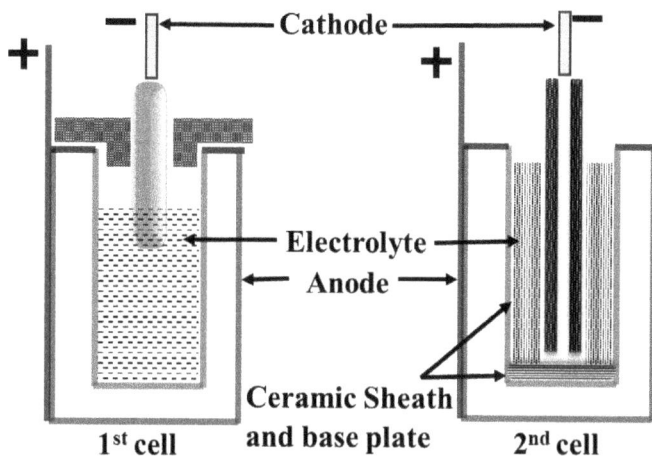

Fig 2.10   Schematic diagram of CNT production by electrolysis. Adapted from Kinloch *et al.* [2003].

solidified salts is washed with copious amounts of distilled water followed by filtration. Fig 2.10 shows two cell designs of CNT synthesis using the electrolysis of liquid NaCl and the second cell is a modified version of the first. The cells are further kept within a heated reactor tube in an inert atmosphere (argon) [Gupta *et al.*, 2014].

2.3.2 *Purification of CNT*

The synthesized CNTs consist of many impurities such as graphite (wrapped up), metal particles, smaller fullerenes, and amorphous carbon. These contaminants influence the required physical and chemical characteristics of nanotubes to a large extent. Purification separates impurities (present in the raw products) from synthesized CNTs and procures desired nanotubes. Purification has a major importance since the discovery of nanotubes. There are different steps of CNT purification such as oxidation, acid treatment, annealing, and ultrasonication. These purification processes eliminate amorphous carbon from the nanotubes, ameliorate surface area, and eliminate different functional groups obstructing the pores. Generally, all these purification steps are employed together in order to enhance the purification and it helps to remove multiple types of contamination simultaneously. The purification techniques are as follows.

2.3.2.1 *Oxidation*

Oxidation of a synthesized nanotube is one of the most commonly used techniques in chemical purification. In this process, CNTs and impurities are oxidized by air or oxygen at a particular temperature. This selective oxidative etching step is based on the fact that carbon impurities such as amorphous carbon and carbon particles have a higher oxidation reaction rate than CNTs and can be eliminated more smoothly than CNTs, so the CNTs are less damaged than the impurities. The impurity oxidation is superior to the impurities that are generally attached to the metal catalyst which act as oxidizing catalysts. Yield and efficiency of the technique rely on various factors such as metal content, temperature, oxidation time, oxidizing agent, and environment.

2.3.2.2 *Acid Treatment*

In this method, nanotubes are refluxed in an acid solution. The concentrated acid treatment removes the carbon nanoparticles, amorphous carbon, and metal catalyst impurities. The surface functionalization of CNTs also happens in the acid treatment process due to the use of various types of acid and acid mixtures such as HCl, HNO_3, and H_2SO_4. The treatment of CNTs with strong oxidizing agents (HNO_3) causes etching of graphitic walls. It is worth noting that during the acid treatment, acid hits the defective sites and the acid molecule intercalates inside the CNT bundles and unfolds the tube walls by oxidative etching that expands the interlayer spacing of nanotubes and the amorphous carbon is removed during oxidative reduction by following equation:

$$C + 4\,HNO_3 = CO_2 + 2H_2O + 4NO \qquad (2.3)$$

It has been studied that the maximum duration for eliminating the majority of the metal particles in nitric reflux is up to 24 h at high temperature.

2.3.2.3 *Annealing and Thermal Treatment*

The high-temperature treatment (600–1500°C) rearranges the nanotubes by consuming the defects and also pyrolyzes the graphitic carbon as well as the short fullerenes present in the CNTs. At high temperature, the catalyst particle (metal) melts and can be easily removed.

2.3.2.4 *Ultrasonication*

The ultrasonication process has been recognized as a distinct and effective technique to remove amorphous carbon impurities. In the presence of an appropriate solvent (ethanol, methanol, dichloromethane, etc.), the nanotubes are treated with high-intensity ultrasonic waves that generate cavitation and are able to dissociate the CNT aggregates formed as a result of van der Waals interactions between the CNTs. These agglomerated nanoparticles are forced to vibrate and get dispersed. The departure of these agglomerated nanoparticles is based on the type of surfactant, solvent, and reagent used. When acids are used in the ultrasonication process, the concentration of acid and time of sonication plays an important role. A concise sonication time can remove the metal catalyst, and an extended sonication time can chemically cut the walls of nanotubes.

2.3.2.5 *Micro-Filtration*

This process depends on the particle size of the nanotubes. In order to free the CNTs from other impurities, a microfiltration membrane with narrow pore size distribution has been used. In this process, CNTs and other amorphous carbon nanoparticles are blocked in a filter of different pore diameters while the other nanoparticles such as metal catalysts, fullerenes, and carbon nanoparticles pass through the filter. The cross-flow filtration is a special form of filtration where filtrate is pumped down at head pressure from a reservoir through a bore of fiber and the solution is reverted in order to be recycled.

2.3.3 *Characterizations of CNTs*

Morphology and microstructural characterization of the CNTs are observed using a reduced number of techniques. It is essential to ascertain the quality and grade of nanotubes using various characterization techniques since different applications require different credentials of nanotubes. The most acceptable characterization techniques are SEM, TEM, STM, Raman spectroscopy, FTIR spectroscopy, X-ray diffraction (XRD), and nuclear magnetic resonance (NMR). Techniques such as TEM and STM characterize CNTs at a distinct level. X-ray photoelectron spectroscopy (XPS) technique determines the chemical structure of CNTs, Raman and FTIR spectroscopies are known as global characterization techniques and are used to determine the presence of functional groups on CNTs and the purity of the sample. Each technique has its own advantages used in conjunction with other techniques.

2.3.3.1 *SEM and TEM*

SEM and TEM are used to reveal the morphology, dimensions, and orientations of CNTs. SEM is used in the preliminary evaluation of the morphology of CNTs but it cannot distinguish metal catalysts and other carbonaceous impurities from CNTs. In order to estimate the metallic content of CNTs, SEM integrated with an energy-dispersive X-ray technique (SEM-EDX) is used. Thus, the SEM technique can dispense information on both the nanotube's morphology and catalyst's impurity content. The TEM technique is used to ascertain the morphological insight, purity, shape, size, and number of layers of nanotubes. TEM can also determine the inner and outer radius of produced CNTs. TEM distinctively provides qualitative information

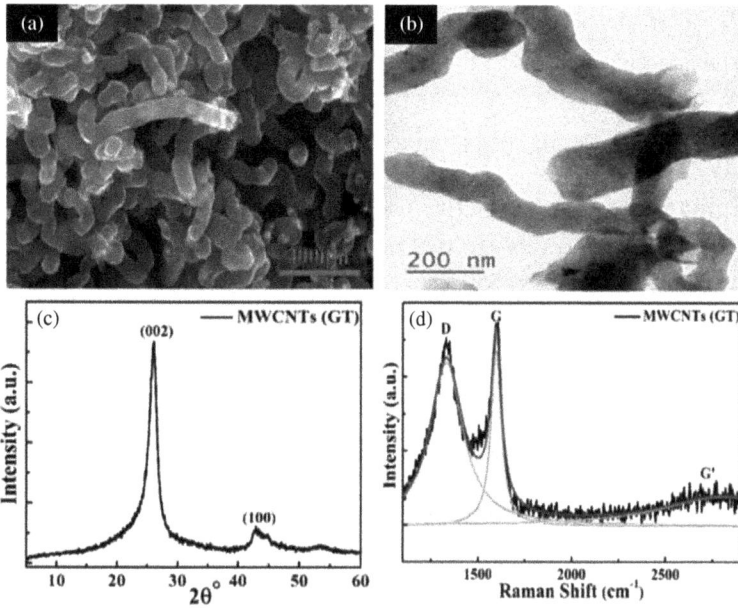

Fig 2.11 (a) SEM image, (b) TEM Image, (c) XRD pattern and (d) Raman spectra of MWCNTs. Reprinted with permission from Mishra *et al.* [2022] Copyright (2023) Springer.

on the size, shape, and structure of trapped metal nanoparticles in a sample. Fig 2.11(a) and (b) shows the SEM and TEM images of CVD-synthesized CNT, respectively.

2.3.3.2 *X-ray Diffraction (XRD)*

This technique is used to find qualitative properties such as crystallinity, interlayer spacing, and impurities. For phase identification of CNTs via XRD, a statistical characterization method is required. Generally, two peak values of CNTs at 26° and 43° are observed by XRD, as shown in Fig 2.11(c). The strongest and sharpest characteristic peak at 26° corresponds to the (002) plane which confirms the presence of a hexagonally symmetric crystalline structure with sp^2 carbon. There is also a small diffraction at $2\theta = 43.0°$ for MWCNTs sample related to graphitic nature. The sample refluxed with nitric acid shows a less intense peak at 53.3°.

2.3.3.3 *Raman Spectroscopy*

It is a quick and nondestructive tool to reveal the remarkable structure and phonon properties of CNTs. All the allotropes of carbon are Raman active with their different positions and relative intensities of bands. The two main characteristic peaks in the Raman spectrum of a purified sample are at 1343 and 1575 cm^{-1}. The band at 1575 cm^{-1} (G-band) is a result of the in-plane vibration of the C–C bond, while the band at 1343 cm^{-1} (D-band) corresponds to the presence of defects in carbon structures. The peak at higher wavenumber side around 2678 cm^{-1} is also observed which indicates the presence of few-layer and less intertube interactions. Fig 2.11(d) shows the Raman spectra of CVD-grown MWCNT.

2.3.4 *Properties of CNTs*

The electronic, mechanical, and structural properties of CNTs are unique due to their one-dimensional structure. Some of the eminent properties of CNTs are as follows.

2.3.4.1 *Electronic Properties*

The most important properties of CNTs depending on their structure are that they can be metallic (such as copper) or semiconducting (such as silicon). According to their chiral vector, the chiral tubes are either semiconducting or metallic whereas the zigzag tubes are metallic. Thus, the molecular structure causes the difference in conducting properties of tubes which gives different band structures and thus a different bandgap. In mathematical terms, this difference in conductivity can be easily given as follows.

The metallic conductance occurs when $n - m$ is a multiple of 3 or an armchair

$$(n = m) \quad \text{or} \quad (n - m = 3i)$$

where n and m are integers that specify the tube's structure and i is an integer. The semiconducting nature is due to $n - m$ not being a multiple of 3 or a zigzag or chiral [Wang & Bao, 2015].

2.3.4.2 *Mechanical Strength*

The mechanical properties of nanotubes such as higher Young's modulus (270–950 GPa) and tensile strength (11–63 GPa) made them one of the strongest materials in nature. Therefore, the nano-tubes are potent and suitable candidates for composite materials with high necessity of anisotropic properties [Yu *et al.*, 2000].

2.3.4.3 *Electrochemical Properties of CNTs*

CNTs exhibit interesting electrochemical properties due to their unique electronic and mechanical structure. CNTs have a very high surface area due to their small diameter and high aspect ratio. CNTs also have excellent electrical conductivity, which makes them suitable for use as electrical contacts and electrodes in various electrochemical devices. They have high mechanical strength, which enables them to withstand mechanical stress and deformation during electrochemical reactions. They also show fast electron transfer kinetics, which facili-tates rapid electron transfer between the electrode and the electrolyte for efficient energy devices. CNTs are chemically stable and can with-stand harsh chemical environments, which makes them suitable for use in various devices [Santos *et al.*, 2019].

2.4 Summary

This chapter provides an overview of various synthesis techniques of carbon nanomaterials (graphene and CNTs) from top-down to bot-tom-up approaches. The top-down approaches are simple to use but suffer from the drawback of improper film shape and sizes, and are appropriate only when bulk crystals are available. These limitations are overcome in the bottom-up approach, where the synthesized films have well-defined shapes, sizes, and chemical compositions. Good synthesis approaches and different derivatives can help in enhancing the performance of these nanomaterials for energy generation and storage applications. In this chapter, we also addressed the characterization of carbon nanostructures and their outstanding electrical, optical, thermal, and mechanical properties.

References

Ajayan, P. M., Lambert, J. M., Bernier, P., Barbedette, L., Colliex, C. & Planeix, J. M. (1993). Growth morphologies during cobalt-catalyzed single-shell carbon nanotube synthesis. *Chem. Phys. Lett.* 215(5), pp. 509–517.

Balandin, A. A., Ghosh, S., Bao, W., Calizo, I., Teweldebrhan, D., Miao, F. & Lau, C. N. (2008). Superior thermal conductivity of single-layer graphene. *Nano Lett.* 8(3), pp. 902–907.

Berger, C., Song, Z., Li, T., Li, X., Ogbazghi, A. Y., Feng, R., Dai, Z., Marchenkov, A. N., Conrad, E. H., First, P. N. & De Heer, W. A. (2004). Ultrathin epitaxial graphite: 2D electron gas properties and a route toward graphene-based nanoelectronics. *J. Phys. Chem. B* 108(52), pp. 19912–19916.

Bhuyan, M. S. A., Uddin, M. N., Islam, M. M., Bipasha, F. A. & Hossain, S. S. (2016). Synthesis of graphene. *Int. Nano Lett.* 6, pp. 65–83.

Blake, P., Hill, E. W., Castro Neto, A. H., Novoselov, K. S., Jiang, D., Yang, R., Booth, T. J. & Geim, A. K. (2007). Making graphene visible. *Appl. Phys. Lett.* 91(6), p. 063124.

Brodie, B. C. (1860). Sur le poids atomique du graphite. *Ann. Chim. Phys.* 59(466), p. 472.

Cano-Márquez, A. G., Rodríguez-Macías, F. J., Campos-Delgado, J., Espinosa-González, C. G., Tristán-López, F., Ramírez-González, D., Cullen, D. A., Smith, D. J., Terrones, M. & Vega-Cantú, Y. I. (2009). Ex-MWNTs: Graphene sheets and ribbons produced by lithium intercalation and exfoliation of carbon nanotubes. *Nano Lett.* 9(4), pp. 1527–1533.

Chen, Z., Lin, Y. M., Rooks, M. J. & Avouris, P. (2007). Graphene nanoribbon electronics. *Physica E Low Dimens. Syst. Nanostruct.* 40(2), pp. 228–232.

Chua, C. K. & Pumera, M. (2014). Chemical reduction of graphene oxide: A synthetic chemistry viewpoint. *Chem. Soc. Rev.* 43(1), pp. 291–312.

Das, A., Chakraborty, B. & Sood, A. K. (2008). Raman spectroscopy of graphene on different substrates and influence of defects. *Bull. Mater. Sci.* 31, pp. 579–584.

Dimitrov, A. T., Tomova, A., Grozdanov, A. & Paunović, P. (2011). Production, purification, characterization, and application of CNTs. in: Nanotechnological basis for advanced sensors. pp. 121–142.

Dimitrov, A. T., Tomova, A., Grozdanov, A., Popovski, O. & Paunović, P. (2013). Electrochemical production, characterization, and application of MWCNTs. *J. Solid State Electrochem.* 17, pp. 399–407.

Guo, T., Nikolaev, P., Rinzler, A. G., Tomanek, D., Colbert, D. T. & Smalley, R. E. (1995). Self-assembly of tubular fullerenes. *J. Phys. Chem.* 99(27), pp. 10694–10697.

Gupta, A., Chen, G., Joshi, P., Tadigadapa, S. & Eklund, P. C. (2006). Raman scattering from high-frequency phonons in supported n-graphene layer films. *Nano Lett.* 6(12), pp. 2667–2673.

Gupta, R. D., Schwandt, C. & Fray, D. J. (2014). Preparation of tin-filled carbon nanotubes and nanoparticles by molten salt electrolysis. *Carbon* 70, pp. 142–148.

Hernandez, Y., Nicolosi, V., Lotya, M., Blighe, F. M., Sun, Z., De, S., McGovern, I. T., Holland, B., Byrne, M., Gun'Ko, Y. K. & Boland, J. J. (2008). High-yield production of graphene by liquid-phase exfoliation of graphite. *Nat. Nanotechnol.* 3(9), pp. 563–568.

Hsu, W. K., Hare, J. P., Terrones, M., Kroto, H. W., Walton, D. R. M. & Harris, P. J. F. (1995). Condensed-phase nanotubes. *Nature* 377(6551), pp. 687–687.

Hummers Jr, W. S. & Offeman, R. E. (1958). Preparation of graphitic oxide. *J. Am. Chem. Soc.* 80(6), pp. 1339–1339.

Iijima, S. (1991). Helical microtubules of graphitic carbon. *Nature* 354 (6348), pp. 56–58.

Iijima, S. & Ichihashi, T. (1993). Single-shell carbon nanotubes of 1-nm diameter. *Nature* 363(6430), pp. 603–605.

Jo, G., Choe, M., Lee, S., Park, W., Kahng, Y. H. & Lee, T. (2012). The application of graphene as electrodes in electrical and optical devices. *Nanotechnol.* 23(11), p. 112001.

Journet, C., Picher, M. & Jourdain, V. (2012). Carbon nanotube synthesis: From large-scale production to atom-by-atom growth. *Nanotechnol.* 23(14), p. 142001.

Juang, Z. Y., Wu, C. Y., Lo, C. W., Chen, W. Y., Huang, C. F., Hwang, J. C., Chen, F. R., Leou, K. C. & Tsai, C. H. (2009). Synthesis of graphene on silicon carbide substrates at low temperature. *Carbon* 47(8), pp. 2026–2031.

Jung, I., Pelton, M., Piner, R., Dikin, D. A., Stankovich, S., Watcharotone, S., Hausner, M. & Ruoff, R. S. (2007). Simple approach for high-contrast optical imaging and characterization of graphene-based sheets. *Nano Lett.* 7(12), pp. 3569–3575.

Kingston, C. T. & Simard, B. (2003). Fabrication of carbon nanotubes. *Anal. Lett.* 36(15), pp. 3119–3145.

Kinloch, I. A., Chen, G. Z., Howes, J., Boothroyd, C., Singh, C., Fray, D. J. & Windle, A. H. (2003). Electrolytic, TEM and Raman studies on the production of carbon nanotubes in molten NaCl. *Carbon* 41(6), pp. 1127–1141.

Kosynkin, D. V., Higginbotham, A. L., Sinitskii, A., Lomeda, J. R., Dimiev, A., Price, B. K. & Tour, J. M. (2009). Longitudinal unzipping of carbon nanotubes to form graphene nanoribbons. *Nature* 458(7240), pp. 872–876.

Krishnamurthy, G. & Namitha, R. (2013). Synthesis of structurally novel carbon micro/nanospheres by low temperature-hydrothermal process. *J. Chil. Chem. Soc.* 58(3), pp. 1930–1933.

Lee, H. C., Liu, W. W., Chai, S. P., Mohamed, A. R., Aziz, A., Khe, C. S., Hidayah, N. M. S. & Hashim, U. (2017). Review of the synthesis, transfer, characterization and growth mechanisms of single and multilayer graphene. *RSC Adv.* 7(26), pp. 15644–15693.

Lee, C., Wei, X., Kysar, J. W. & Hone, J. (2008). Measurement of the elastic properties and intrinsic strength of monolayer graphene. *Science* 321(5887), pp. 385–388.

Lee, J., Zheng, X., Roberts, R. C. & Feng, P. X. L. (2015). Scanning electron microscopy characterization of structural features in suspended and non-suspended graphene by customized CVD growth. *Diamond Relat. Mater.* 54, pp. 64–73.

Li, Y., Chen, Q., Xu, K., Kaneko, T. & Hatakeyama, R. (2013). Synthesis of graphene nanosheets from petroleum asphalt by pulsed arc discharge in water. *Chem. Eng. J.* 215, pp. 45–49.

Li, X., Zhu, Y., Cai, W., Borysiak, M., Han, B., Chen, D., Piner, R. D., Colombo, L. & Ruoff, R. S. (2009). Transfer of large-area graphene films for high-performance transparent conductive electrodes. *Nano Lett.* 9(12), pp. 4359–4363.

Lu, X., Yu, M., Huang, H. & Ruoff, R. S. (1999). Tailoring graphite with the goal of achieving single sheets. *Nanotechnol.* 10(3), p. 269.

Manafi, S., Nadali, H. & Irani, H. R. (2008). Low temperature synthesis of multi-walled carbon nanotubes via a sonochemical/hydrothermal method. *Mater. Lett.* 62(26), pp. 4175–4176.

Marcano, D. C., Kosynkin, D. V., Berlin, J. M., Sinitskii, A., Sun, Z., Slesarev, A., Alemany, L. B., Lu, W. & Tour, J. M. (2010). Improved synthesis of graphene oxide. *ACS Nano.* 4(8), pp. 4806–4814.

Meyer, J. C., Kisielowski, C., Erni, R., Rossell, M. D., Crommie, M. F. & Zettl, A. (2008). Direct imaging of lattice atoms and topological defects in graphene membranes. *Nano Lett.* 8(11), pp. 3582–3586.

Mishra, S., Jaiswal, S. S. & Mishra, A. K. (2022). Multiwalled carbon nanotubes grown over green iron nanocatalyst as electrode for hydrogen-producing electrochemical cell. *J. Mater. Sci.: Mater. Electron.* 33(11), pp. 8702–8710.

Nair, R. R., Blake, P., Grigorenko, A. N., Novoselov, K. S., Booth, T. J., Stauber, T., Peres, N. M. & Geim, A. K. (2008). Fine structure constant defines visual transparency of graphene. *Science* 320(5881), pp. 1308–1308.

Ni, Z. H., Wang, H. M., Kasim, J., Fan, H. M., Yu, T., Wu, Y. H., Feng, Y. P. & Shen, Z. X. (2007). Graphene thickness determination using reflection and contrast spectroscopy. *Nano Lett.* 7(9), pp. 2758–2763.

Novoselov, K. S., Geim, A. K., Morozov, S. V., Jiang, D. E., Zhang, Y., Dubonos, S. V., Grigorieva, I. V. & Firsov, A. A. (2004). Electric field effect in atomically thin carbon films. *Science* 306(5696), pp. 666–669.

Nozaki, T. & Okazaki, K. (2008). Carbon nanotube synthesis in atmospheric pressure glow discharge: A review. *Plasma Processes Polym.* 5(4), pp. 300–321.

Nozaki, T., Yoshida, S., Karatsu, T. & Okazaki, K. (2011). Atmospheric-pressure plasma synthesis of carbon nanotubes. *J. Phys. D: Appl. Phys.* 44(17), p. 174007.

Paredes, J. I., Villar-Rodil, S., Solís-Fernández, P., Martínez-Alonso, A. & Tascon, J. M. D. (2009). Atomic force and scanning tunneling microscopy imaging of graphene nanosheets derived from graphite oxide. *Langmuir* 25(10), pp. 5957–5968.

Rao, C. N. R., Subrahmanyam, K. S., Matte, H. R., Abdulhakeem, B., Govindaraj, A., Das, B., Kumar, P., Ghosh, A. & Late, D. J. (2010). A study of the synthetic methods and properties of graphenes. *Sci Technol Adv Mater.*

Santos, C., Senokos, E., Fernández-Toribio, J. C., Ridruejo, Á., Marcilla, R. & Vilatela, J. J. (2019). Pore structure and electrochemical properties of CNT-based electrodes studied by in situ small/wide angle X-ray scattering. *J. Mater. Chem. A* 7(10), pp. 5305–5314.

Smith, A. T., LaChance, A. M., Zeng, S., Liu, B. & Sun, L. (2019). Synthesis, properties, and applications of graphene oxide/reduced graphene oxide and their nanocomposites. *Nano Mater. Sci.* 1(1), pp. 31–47.

Smith, S. C. & Rodrigues, D. F. (2015). Carbon-based nanomaterials for removal of chemical and biological contaminants from water: A review of mechanisms and applications. *Carbon* 91, pp. 122–143.

Staudenmaier, L. (1898). Verfahren zur darstellung der graphitsäure. *Ber. Dtsch. Chem. Ges.* 31(2), pp. 1481–1487.

Subrahmanyam, K. S., Panchakarla, L. S., Govindaraj, A. & Rao, C. N. R. (2009). Simple method of preparing graphene flakes by an arc-discharge method. *J. Phys. Chem. C* 113(11), pp. 4257–4259.

Takahashi, K., Yamada, K., Kato, H., Hibino, H. & Homma, Y. (2012). In situ scanning electron microscopy of graphene growth on polycrystalline Ni substrate. *Surf. Sci.* 606(7–8), pp. 728–732.

Tanaka, H., Arima, R., Fukumori, M., Tanaka, D., Negishi, R., Kobayashi, Y., Kasai, S., Yamada, T. K. & Ogawa, T. (2015). Method for controlling electrical properties of single-layer graphene nanoribbons via adsorbed planar molecular nanoparticles. *Sci. Rep.* 5(1), pp. 12341.

Tasis, D., Tagmatarchis, N., Bianco, A. & Prato, M. (2006). Chemistry of carbon nanotubes. *Chem. Rev.* 106(3), pp. 1105–1136.

Thomsen, C. & Reich, S. (2000). Double resonant Raman scattering in graphite. *Phys. Rev. Lett.* 85(24), p. 5214.

Tu, Z., Liu, Z., Li, Y., Yang, F., Zhang, L., Zhao, Z., Xu, C., Wu, S., Liu, H., Yang, H. & Richard, P. (2014). Controllable growth of 1–7 layers of graphene by chemical vapour deposition. *Carbon* 73, pp. 252–258. 4

Unrau, C. J., Axelbaum, R. L. & Lo, C. S. (2010). High-yield growth of carbon nanotubes on composite Fe/Si/O nanoparticle catalysts: A Car–Parrinello molecular dynamics and experimental study. *J. Phys. Chem. C* 114(23), pp. 10430–10435.

Van Khai, T., Kwak, D. S., Kwon, Y. J., Cho, H. Y., Huan, T. N., Chung, H., Ham, H., Lee, C., Van Dan, N., Tung, N. T. & Kim, H. W. (2013). Direct production of highly conductive graphene with a low oxygen content by a microwave-assisted solvothermal method. *Chem. Eng. J.* 232, pp. 346–355.

Viculis, L. M., Mack, J. J., Mayer, O. M., Hahn, H. T. & Kaner, R. B. (2005). Intercalation and exfoliation routes to graphite nanoplatelets. *J. Mater. Chem.* 15(9), pp. 974–978.

Wang, H. & Bao, Z. (2015). Conjugated polymer sorting of semiconducting carbon nanotubes and their electronic applications. *Nano Today* 10(6), pp. 737–758.

Wang, Z., Li, N., Shi, Z. & Gu, Z. (2010). Low-cost and large-scale synthesis of graphene nanosheets by arc discharge in air. *Nanotechnol.* 21(17), p. 175602.

Wang, J., Zhu, M., Outlaw, R. A., Zhao, X., Manos, D. M. & Holloway, B. C. (2004). Synthesis of carbon nanosheets by inductively coupled radio-frequency plasma enhanced chemical vapor deposition. *Carbon* 42(14), pp. 2867–2872.

Wong, S. L., Huang, H., Chen, W. & Wee, A. T. (2012). STM studies of epitaxial graphene. *MRS Bull.* 37(12), pp. 1195–1202.

Yoon, D., Moon, H., Cheong, H., Choi, J. S., Choi, J. A. & Park, B. H. (2009). Variations in the Raman spectrum as a function of the number of graphene layers. *J. Korean Phys. Soc.* 55(3), pp. 1299–1303.

Yu, Y., Cui, C., Qian, W., Xie, Q., Zheng, C., Kong, C. & Wei, F. (2013). Carbon nanotube production and application in energy storage. *Asia-Pac. J. Chem. Eng.* 8(2), pp. 234–245.

Yu, M. F., Lourie, O., Dyer, M. J., Moloni, K., Kelly, T. F. & Ruoff, R. S. (2000). Strength and breaking mechanism of multiwalled carbon nanotubes under tensile load. *Science* 287(5453), pp. 637–640.

Yuan, G. D., Zhang, W. J., Yang, Y., Tang, Y. B., Li, Y. Q., Wang, J. X., Meng, X. M., He, Z. B., Wu, C. M. L., Bello, I. & Lee, C. S. (2009). Graphene sheets via microwave chemical vapor deposition. *Chem. Phys. Lett.* 467(4–6), pp. 361–364.

Zhang, Y., Small, J. P., Pontius, W. V. & Kim, P. (2005). Fabrication and electric-field-dependent transport measurements of mesoscopic graphite devices. *Appl. Phys. Lett.* 86(7), p. 073104.

Zhao, X., Ohkohchi, M., Wang, M., Iijima, S., Ichihashi, T. & Ando, Y. (1997). Preparation of high-grade carbon nanotubes by hydrogen arc discharge. *Carbon* 35(6), pp. 775–781.

Zhu, Y., Murali, S., Cai, W., Li, X., Suk, J. W., Potts, J. R. & Ruoff, R. S. (2010). Graphene and graphene oxide: Synthesis, properties, and applications. *Adv. Mater.* 22(35), pp. 3906–3924.

Chapter 3

Synthesis, Characterization, and Properties of TMDs Nanostructures

Ankita Singh and Ashish Kumar Mishra

*School of Materials Science and Technology, Indian Institute of
Technology (Banaras Hindu University), Varanasi 221005, India
Email: akmishra.mst@iitbhu.ac.in*

Abstract

In the last two decades, vibrant worldwide research on two-dimensional (2D) materials triggered significant attention owing to their outstanding electrical, mechanical, and optical properties. Among 2D materials, transition metal dichalcogenides (TMDs) such as MoS_2, $MoSe_2$, and WS_2 are a large family of layered materials that exhibits peculiar and fascinating properties leading to unlimited potential in electronics, optoelectronics, and energy storage applications. The characteristics and properties of these TMDs determine their utilization in energy generation and storage applications. This chapter intends to discuss the various synthesis routes, mainly categorized into two types of approaches, namely top-down and bottom-up. Furthermore, the characterization techniques along with some of the key properties of TMD nanostructures are also discussed.

3.1 Introduction

Recently, a class of material called two-dimensional transition metal dichalcogenides (2D-TMDs) has been investigated and achieved intense attraction in the energy research community. TMDs play a vital character in improving the performances of energy technologies, owing to the large surface area, thin structures, enhanced kinetics, etc. for electrochemical activity. Its layered structure also facilitates quick intercalation and de-intercalation of ions due to weak interlayer bonding and has enhanced the performance of supercapacitors and secondary batteries. MX_2 such as MoS_2, WS_2, $MoSe_2$, and WSe_2 are the members of a large class of TMD materials. It possesses a graphene-like layered structure with the repeat unit X-M-X. As an example, single-layer MoS_2 is three atoms thick having two hexagonal planes of S atoms separated by the plane of Mo atoms with strong covalent bond within each layer and weak van der Waal interaction between the layers with an interlayer spacing of ~0.615 nm. Based on the stacking of this repeat unit along the c-axis, MoS_2 exhibits three different polytypes: 1T, 2H, and 3R, with the point groups D_{6d}, D_{6h}, and C_{3v}, respectively [Wang *et al.*, 2012]. The first digit represents the layer number and the second alphabet shows the crystallographic structure (i.e., T for trigonal, H for hexagonal, and R for rhombohedral structure). The 2H phase (trigonal prismatic coordination) exhibits hexagonal symmetry having two S-Mo-S layers per repeat unit, while the 3R phase (trigonal prismatic coordination) shows rhombohedral symmetry having three S-Mo-S layers per repeat unit. In the 1T phase (octahedral coordination), the upper triangle is the inversion of the bottom triangle and shows tetragonal symmetry having one S-Mo-S layer per repeat unit. The 1T and 3R phases are metastable structures showing conductive characteristics, while the 2H phase is a stable semiconducting structure with a bandgap in the range of 1.23–1.88 eV. The electronic band structure is the key determinant for the application of MoS_2.

Like MoS_2, $MoSe_2$ is also an important layered semiconductor (Se-Mo-Se) having a narrow bandgap (1.09 to 1.57 eV) and provides a better option for energy storage applications. $MoSe_2$ has an interlayer spacing of 0.646 nm, more appreciable than that of MoS_2

(0.615 nm) and graphite (0.335 nm), and hence can be used as a good choice for electrode material in energy devices. Researchers have also tried to increase the electrochemical performance of $MoSe_2$ by designing a hybrid nanostructure with a carbon matrix. In addition to most studied MoS_2 and $MoSe_2$, other TMDs have also captured attention for energy storage applications [Singh *et al.*, 2023]. This chapter intends to briefly discuss different preparation methods and characterization of a few TMDs (mainly sulfides and selenides based) along with their key properties.

3.2 Transition Metal Disulfides (MS$_2$)

The transition metal disulfide family includes compounds such as MoS_2, WS_2, SnS_2, TaS_2, and NbS_2. Among all, MoS_2 is the most widely investigated TMD nanomaterial for energy generation and storage applications. Various synthesis methods, characterizations, and properties of MoS_2 are discussed in the following subsections.

3.2.1 *Synthesis of MoS$_2$*

Various morphologies of MoS_2 such as 0D (quantum dot, nanoplatelets), 1D (nanorods, nanowires), 2D (nanosheets, nanoribbons, nanoflakes), and 3D (nanoflowers) can be obtained using different synthesis techniques. Modifications in the synthesis route can yield high-quality, uniform, and large-area MoS_2 with controlled thickness. Synthesis of MoS_2 is mainly categorized based on two approaches, namely top-down and bottom-up. The former approach includes methods such as exfoliation (mechanical and chemical), wet chemical synthesis, arc discharge, and sputtering, while the latter approach includes methods such as chemical vapor deposition (CVD) (thermal and plasma enhanced), physical vapor deposition, and atomic layer deposition. This chapter discusses the most frequently used methods.

3.2.1.1 *Mechanical Exfoliation*

High-quality monolayer graphene, exfoliated from graphite, inspired the synthesis of other 2D materials by the simple "Scotch tape method" using external stress. In micromechanical exfoliation, bulk MoS_2 is attached and peeled off using scotch tape. Then it is pressed onto the substrate and released with the deposition of random shapes and sizes of varying layer numbers. The disadvantage of this method is its low production yield and lack of controllability over flake thickness and its size. Nanomechanical cleavage has been developed to overcome the problem of randomness in layer numbers. In this method, a sharp tungsten tip ($\sim$10 nm) is attached to a piezoelectric motor and made contact with the crystalline edges of bulk MoS_2 for the selective layer number cleavage [Tang *et al.*, 2014]. The MoS_2 crystal is hooked up to a support with an edge-on orientation. After contact, the tip is bent at an angle so that the monolayer gets attached to the probe, as shown in Fig 3.1(a). By moving the tip away, the flake gets stretched and separated from the crystal. Gacem *et al.* [2012] proposed an alternative method called anodic bonding (Fig 3.1(b)) that assures higher yield and easy transferability on an arbitrary substrate. In this method, bulk MoS_2 is placed on top of Pyrex glass and heated at 130–200°C under a high voltage of 200–1500 V to form a bonding with the substrate. Then with the help of adhesive tape, it is mechanically peeled off for producing a thin film.

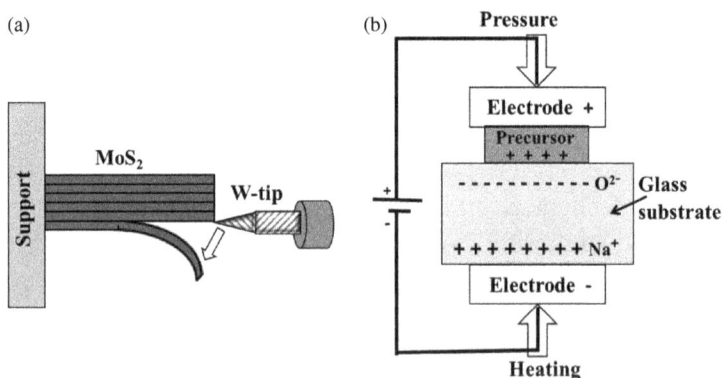

Fig 3.1   (a) Schematic showing the setup for nanomechanical cleavage of MoS_2 using the tungsten tip. Adapted from Tang *et al.* [2014]. (b) Schematic showing the anodic bonding. Adapted from Gacem *et al.* [2012].

3.2.1.2 *Chemical Exfoliation*

A low yield problem was overcome by another alternative method called chemical/liquid exfoliation. It includes two methods: ion intercalation and solvent exfoliation. The former method was first developed by Joensen *et al.* [1986]. In this method, Li is intercalated between layers of MoS_2 and Li intercalated MoS_2 reacts with water, forming H_2 gas whose expansion leads to the separation of layers of MoS_2. Instead of water, the sample can also be exfoliated in methanol, ethanol, and isopropanol with rapid heating (~600°C). The drawback of this method is its long reaction time and lack of control over the intercalation process. Incomplete insertion of Li results in low yield, and over-intercalation leads to decomposition into Li_2S. It also results in the transformation of 2H into 1T phase, which is undesirable for electronic applications. Therefore, further annealing is required to restore the phase.

Zeng *et al.* [2011] developed an electrochemical lithiation process, for controlling intercalation. They carried out the process in a test cell with Li foil as an anode and MoS_2 as a cathode. The discharge process provides complete monitoring of the Li intercalation. Despite these alternatives, Li intercalation has the demerit of losing its semiconducting properties. Moreover, a difficulty arises in the complete removal of Li that generates a doping effect. Therefore, another guest species in the electrochemical route need to be introduced to produce MoS_2 nanosheets. Liu *et al.* [2014] investigated the intercalation of SO_4^{2-} anions in a bulk MoS_2 crystal. Fig 3.2 displays the schematic experimental setup of the same. In this method, a voltage of +10 V is applied between MoS_2 (working electrode) and Pt wires (counter electrode) in a 0.5-M Na_2SO_4 solution for 0.5–2 h, resulting in the exfoliation of the MoS_2 crystal. In this process, ·OH radicals, ·O radicals, and SO_4^{2-} anions intercalate between the layers which weakens the strength of van der Waals' interactions. Oxidation of radicals and anions produces O_2 and SO_2 gas bubbles that help in the detachment of flakes of MoS_2 from the bulk. In solvent-based exfoliation (also named as Coleman method), bulk MoS_2 is immersed in solvents such as *N*-methyl-2-pyrrolidone (NMP) and isopropanol (IPA) to minimize the energy of

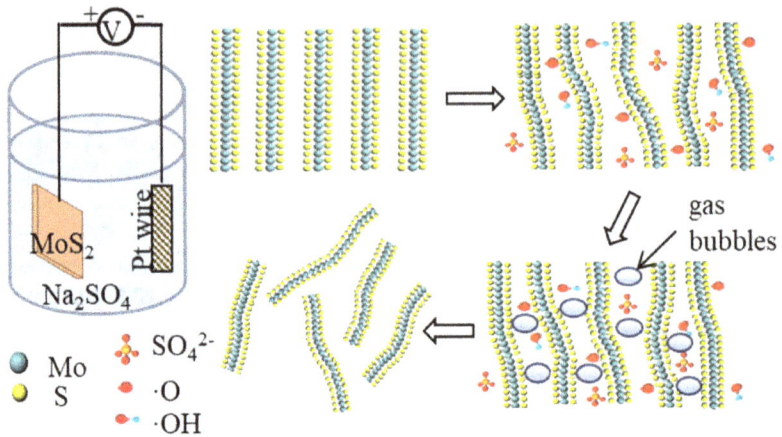

Fig 3.2   Schematic showing the synthesis of 2D MoS_2 nanosheets by elec-trochemical intercalation and exfoliation by ·OH radical, ·O radical and SO_4^{2-} anions. Adapted from Liu *et al.* [2014].

exfoliation, followed by sonication [Coleman *et al.*, 2011], which results in the formation of monolayer to few-layer MoS_2 nanosheets and can be scaled up for large quantity exfoliated film. This method can also be used to produce hybrid films exhibiting enhanced properties by blending with the suspension of other nanomaterials.

3.2.1.3 *Chemical Vapor Deposition*

It is one of the compatible methods for the fabrication of large-area uniform MoS_2. The CVD growth setup is shown in Fig 3.3. Chemical reactions occur during growth, resulting in the deposition of a thin film on the substrate. At reaction temperature, Mo precursor, such as MoO_3, $(NH_4)_2MoS_4$, or Mo, deposits on the substrate followed by sulfurization. Wang *et al.* [2013] fabricated a highly crystalline MoS_2 film using MoO_2 microcrystals. MoO_3 powder was first evaporated and reduced by S vapor to form MoO_2 that nucleated on SiO_2/Si substrate. Then, annealing was done in the presence of S vapor at 850–950°C for 0.5–6 h using Ar gas. Annealing sulfurizes the MoO_2 microplates, producing MoS_2 in rhomboid shape with different layer

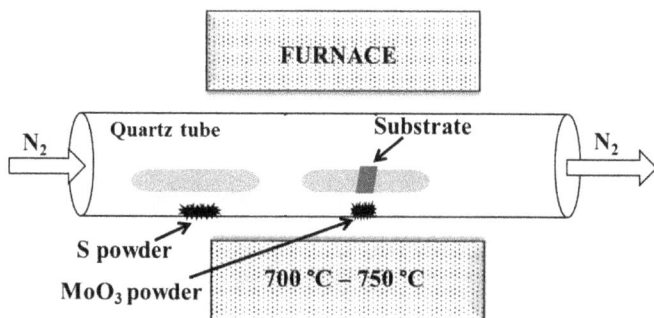

Fig 3.3 Schematic illustrating the CVD setup for the synthesis of MoS₂ film using MoO₃ and S as precursors.

numbers and domain size of ~10 μm. Using the same precursor but different growth conditions, edge–terminated vertically oriented (V-type) few-layer MoS$_2$ sheets were synthesized, which is useful in optoelectronic applications [Majee *et al.*, 2020]. The process involves the vaporization of MoO$_3$ powder using S at 750°C for 10 min in a nitrogen environment, resulting in the growth of V-type MoS$_2$ film. A simple and straightforward method reported is by vapor–solid growth mechanism using MoS$_2$ powder as a precursor [Wu *et al.*, 2013]. In this, MoS$_2$ powder is kept in the middle of the furnace with an insulating substrate (SiO$_2$/Si, sapphire, or normal glass) placed downstream. The reaction is implemented at ~ 900°C for 15–20 min in an Ar environment, producing high-quality triangular 1L MoS$_2$. Further, a new self-limiting CVD technique is developed using MoCl$_5$ and S as precursors [Yu *et al.*, 2013]. The MoCl$_5$ and S powder were kept in the middle and upstream positions of the furnace, respectively, with the substrate (SiO$_2$, sapphire, or graphite) at downstream. The furnace was raised at 850°C for 10 min in the presence of an Ar environment resulting in the successful growth of monolayer or few-layer MoS$_2$ films of a centimeter size. Here, layer number is controlled by the amount of MoCl$_5$ precursor. In another CVD process, elemental Mo is used for growing large-area single-crystal MoS$_2$ [Laskar *et al.*, 2013]. First, e-beam evaporation is used to deposit a thin layer of Mo metal (50 Å) on the substrate. At high temperatures, sulfur vaporizes and reacts with Mo on the substrate in an inert atmosphere. The obtained film depends upon the area coverage of Mo, the type of substrate and its thickness, the reaction temperature, and the amount of sulfur used.

The above-mentioned CVD synthesis methods require a temperature of 600–800°C or even greater than 1000°C during sulfurization of Mo film. Such high temperature limits the CVD synthesis process directly on the device substrate. TMD is instead transferred from the synthesized substrate to a target substrate, which ultimately adds residues that affect the device's performance. Plasma-assisted chemical vapor deposition (PECVD) provides a way to synthesize MoS_2 at a lower growth temperature of 150–300°C with direct synthesis on an arbitrary substrate, even a plastic substrate [Ahn *et al.*, 2015]. A low temperature of 400°C produces uniform few-layer MoS_2 film by employing vapor phase MoO_3 with H_2S plasma on an arbitrary substrate by PECVD [Campbell *et al.*, 2017]. Metal-organic chemical vapor deposition (MOCVD) is another type of CVD that requires organometallic precursors for the deposition of highly crystalline semiconducting thin films. Kang *et al.* [2015] reported the growth of MoS_2 using gaseous precursor molybdenum hexacarbonyl ($Mo(CO)_6$) and diethyl sulfide (($C_2H_5)_2S$) on a 4-inch substrate (fused silica or SiO_2/Si) by MOCVD. A growth time of ~26 h at 550°C produces monolayer MoS_2 on the substrate.

3.2.1.3.1 Transfer mechanism

For practical applications, it is crucial to transfer the as-grown MoS_2 film from the previous substrate to the desired one. Various transfer methods were developed, but it still remains a challenging task to get damage-free film. As for graphene, a similar method was adapted for transferring MoS_2, that is, wet-etching process. For this, oxide etchants such as HF, HCl, HNO_3, and NaOH are used, which degrade the film. Phan *et al.* [2017] developed an ultraclean and direct one-step transfer method of MoS_2 grown over SiO_2 without contaminating the film. They transferred the film to a PET (polyethylene terephthalate) substrate by applying epoxy glue. Initially, the

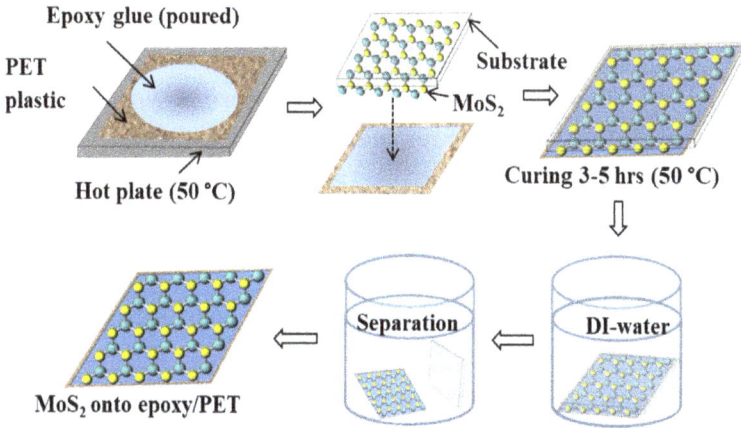

Fig 3.4 Schematic showing the transfer process of MoS_2 using epoxy glue. Adapted from Phan *et al.* [2017].

epoxy glue was spread on the PET substrate heated at 50°C. Then, the MoS_2 on SiO_2/Si substrate with face down is placed on the epoxy glue. Finally, the substrate/MoS_2/epoxy/PET stack is maintained at 50°C for 3–5 h and placed in the DI water leading to the detachment of substrate from MoS_2/epoxy/PET. Fig 3.4 shows the schematic of the step-by-step transfer process using epoxy glue. The advantage of this method is that no residues are left on the surface and the substrate can be recycled for the next use. Researchers are still looking for alternative methods that are efficient and cost-effective for transferring large-area films without damaging and contaminating the film.

3.2.1.4 *Wet Chemical Synthesis*

It is a bottom-up approach for high-yield production of nanomaterials with different morphology. The growth mechanism consists of nucleation in a solution medium followed by growth, which results in a particular architecture depending on the growth condition. It is classified into three types: hydrothermal, solvothermal, and colloidal syntheses. The colloidal route is carried out in a three-neck round-bottom flask in an inert atmosphere at a higher temperature and shorter growth time compared to hydro/solvothermal method.

Hydrothermal and solvothermal syntheses are alike in several aspects such as precursors and autoclave used and the chemical reactions. But the difference between the two is that hydrothermal synthesis involves chemical reactions in an aqueous solution, while solvothermal synthesis is accomplished in a nonaqueous medium (organic solvent such as DMF, pyridine, or octylamine). In both of these methods, Mo and S precursors are first made to dissolve in a suitable solvent by stirring or sonication. Then it is made to react at a moderate temperature (150–250°C) and a high pressure for several hours (10–48 h) in a tightly sealed Teflon-lined stainless steel container called an autoclave. Finally, after cooling, it is washed and left for overnight drying resulting in a few-layer MoS_2 structure. To enhance the purity and crystallinity, it is often annealed at high temperatures. Synthesized MoS_2 may vary in terms of morphology, phase formed, dimension of particle, and crystallinity depending on the type of precursors chosen and their concentration, growth time, growth temperature, and pH value. The 1T and 2H phases are formed based on the growth temperature. A major challenge in this method is the selectivity and limited usage of Mo and S precursors. Some of the reported precursors for Mo are molybdenum trioxide (MoO_3), molybdenum pentachloride ($MoCl_5$), sodium molybdate (Na_2MoO_4), ammonium heptamolybdate ((NH_4)$_6Mo_7O_{24}$), etc., and S precursors include thiourea (NH_2CSNH_2), L-cysteine ($C_3H_7NO_2S$), dibenzyl disulfide ($C_{14}H_{14}S_2$), potassium thiocyanate (KSCN), thioacetamide (C_2H_5NS), etc. The single precursor used for both Mo and S is ammonium tetrathiomolybdate ((NH_4)$_2MoS_4$). Adding NaOH and HCl governs the pH value, while some reducing agents such as N_2H_4 and $NaBH_4$ are also used. Synthesized MoS_2 may have a variety of morphologies such as nano-sized sheets, flakes, flowers, dots, rods, spheres, and core-shell structures, with nanosheets being more popular among all.

Ren *et al.* [2015] synthesized monolayer MoS_2 quantum dots by hydrothermal method using Na_2MoO_4 and $C_{14}H_{14}S_2$ as Mo and S precursors, respectively. The reaction was executed at 220°C for 18 h in an autoclave. Fig 3.5(a) shows the schematic of the synthesis process. Chung *et al.* [2014] synthesized MoS_2 nanospheres and nanosheets using the hydrothermal approach with Na_2MoO_4 and

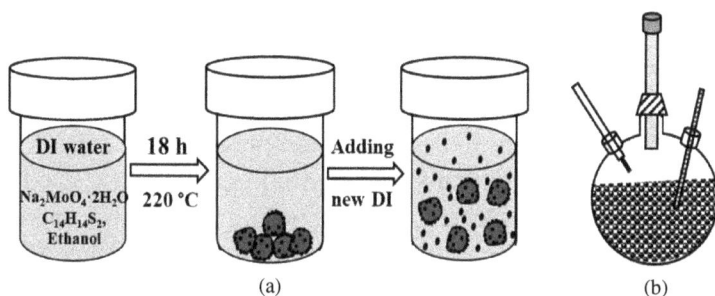

Fig 3.5 Schematic illustrating the (a) hydrothermal method for the synthesis of MoS$_2$ QD and (b) colloidal synthesis method for MoS$_2$ nanosheets. Adapted from Ren *et al.* [2015].

L-cysteine as Mo and S sources, respectively. After adding HCl drop by drop (till pH < 1), the reaction is performed at 220°C for 36 h. For synthesizing MoS$_2$ nanosheets, they used the same procedure, by replacing S precursor (L-cysteine) with thiourea. Li *et al.* [2005] synthesized MoS$_2$ nanowires and nanoribbons using the precursor MoO$_x$. This precursor is exposed with H$_2$S at a temperature of 500–700°C, producing MoS$_2$ nanowires of the 2H phase and above 800°C, forming MoS$_2$ nanoribbons of the 3R phase. Another approach of the wet chemical method to be discussed is the colloidal synthesis method, as shown in Fig 3.5(b). It enables the growth of high-crystalline monolayer nanostructures with monodispersity as well as control over the edges. The growth mechanism is associated with quick nucleation followed by growth. If the precursor's concentration falls below the critical concentration, which is essential for the initiation of nucleation, nucleation ceases after a short duration. Lin *et al.* [2015] designed a colloidal route for the synthesis of MoS$_2$ quantum dots (QD). A mixture of (NH$_4$)$_2$MoS$_4$, oleic acid (OA), oleyl amine (OLA), and 1-octadecene (ODE) is heated at 120°C for 2.5 h under continuous stirring to get a homogeneous solution. Later the prepared solution is raised to 250°C for 3 h, resulting in the formation of MoS$_2$ QD. The OLA reduces (NH$_4$)$_2$MoS$_4$ and also acts as a stabilizing agent that controls the size and property of MoS$_2$. Zhou *et al.* [2016] described the synthesis of single- and multilayer MoS$_2$

nanosheets through controllable colloidal routes. A mixture of molybdenum(II) acetate dimer ($[Mo(Ac)_2]_2$), stearic acid (SA), oleyl amine (OAm), and trioctylphosphine oxide (TOPO) was heated in a three-neck round bottom flask at a temperature of 330°C under nitrogen flow. Various lateral sizes of MoS_2 can be synthesized ranging from 8 to 25 nm by changing the concentration ratio of Mo and S precursors, whereas a multi-injection approach can be used for obtaining MoS_2 with different layer numbers.

3.2.2 *Characterization Techniques of MoS₂*

Various characterization techniques ranging from microscopic to spectroscopic are employed to examine the structural, vibrational, optical, and thermal properties of as-grown monolayer and multi-layer stacks of MoS_2. Some of them are discussed in the following subsections.

3.2.2.1 *Raman Spectroscopy*

It is a characteristic tool for the study of vibrational modes by analyzing the inelastically scattered light (laser) to know the fingerprint of the material at the molecular level. Raman spectroscopy was used as an important characteristic tool for 2D materials to detect the layer number and their phase. The lattice vibrations of $2H$-MoS_2 are schematically shown in Fig 3.6(a). There are four Raman-active modes (E_{2g}^2, E_{1g}, E_{2g}^1, and A_{1g}) and two IR-active modes (E_{1u}^1 and A_{2u}^1). To define the crystal structure of MoS_2, E_{2g}^1, and A_{1g} modes need to be inspected. The E_{2g}^2 peak was first experimentally detected by Verble *et al.* [1972] using a triple Raman spectrometer, observed at 33.7 ± 1 cm⁻¹. This low-frequency peak is usually not observed because of the appearance of strong Rayleigh scattering. The E_{1g} mode (287 cm⁻¹) cannot be detected because of the laser polarization chosen and the random orientation of the crystallites in the sample [Stacy *et al.*, 1985]. Fig 3.6(b, c) shows the optical image and respective Raman spectra of different layered MoS_2 flakes on SiO_2/Si substrates grown

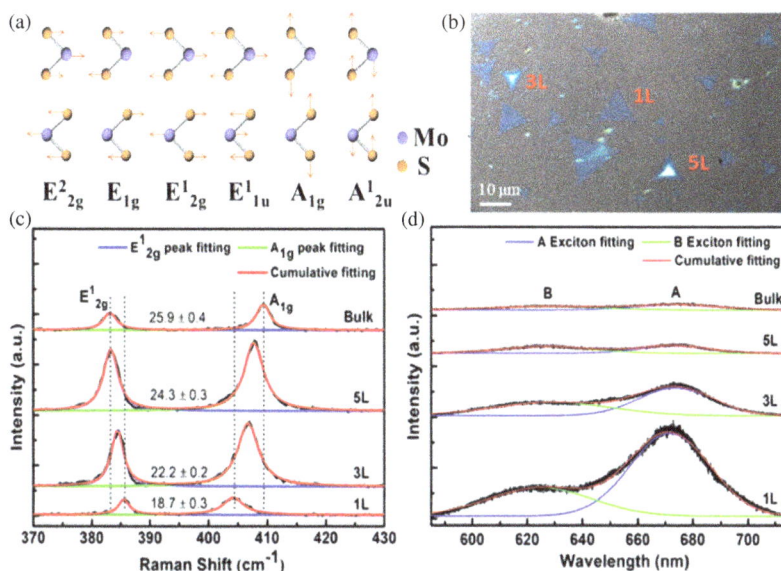

Fig 3.6 (a) Schematic demonstrating the atomic displacements of the four Raman-active modes (E_{2g}^2, E_{1g}, E_{2g}^1, A_{1g}) and two IR-active modes (E_{1u}^1, A_{2u}^1) in the unit cell of the bulk MoS_2 crystal in the [1000] direction. (b) Optical image of different layered CVD-grown triangular MoS_2 nanostructures on SiO_2/Si substrate. (c) Raman spectra (Lorentzian fitted) and (d) PL spectra (Gaussian fitted) of different layers of MoS_2. Fig 3.6 (b–d) Reprinted with permission from Singh *et al.* [2023]. Copyright (2023) American Chemical Society.

by the CVD method. The E_{2g}^1 mode arises due to the in-plane (shear) vibration of S atoms relative to the Mo atoms and is observed at ~383 cm⁻¹, whereas A_{1g} mode is a result of the out-of-plane (breathing) vibration of S atoms and is observed at ~406 cm⁻¹. The E_{2g}^1 peak undergoes red-shift, whereas the A_{1g} peak exhibits blue-shift with increasing layer number. This anomalous frequency trend is attributed to the Coulomb interaction and the induced charge that is developed when layers get stacked. The peak separation between these two typical peaks determines the number of layers in MoS_2. For monolayer, this difference is ~19 cm⁻¹, and for bulk, it is ~26 cm⁻¹ [Singh *et al.*, 2023].

3.2.2.2 *Photoluminescence (PL) Spectroscopy*

Moving from bulk to single-layer MoS_2, indirect to direct energy gap transformation occurs that results in the detection of PL. For bulk MoS_2, the indirect bandgap emerges from the maxima of the valence band (VB) at Γ point to the minima of the conduction band (CB) situated in between Γ and K points; whereas for monolayer MoS_2, the transition from the maxima of the VB to the minima of the CB is observed at same K point, resulting in direct bandgap. Thus, the relaxation of excitons at the K point of monolayer MoS_2 emits photons and these excitonic transitions are responsible for the PL signal. Two excitonic peaks such as A and B excitons are observed in the PL spectrum at 627 and 677 nm, respectively. It is because of the *d–d* transitions from the filled d_x orbital to the degenerated d_{xy} and $d_{x^2-y^2}$ orbitals, split due to spin–orbit coupling [Tanaka *et al.*, 1978], thus permitting two optically active transitions. When film thickness increases from monolayer to bilayer, a small red shift (~20 meV) in the PL energy is observed. This less sensitivity of the direct bandgap to the confinement effect is a result of large electron and hole mass around the K point. The different layered samples can be differentiated based on their PL spectra, as shown in Fig 3.6(d). The peak intensity decreases and diminishes with increasing layer number of MoS_2.

3.2.2.3 *Microscopic Characterization*

Microscopic characterization tools include scanning electron microscopy (SEM), atomic force microscopy (AFM), and transmission electron microscopy (TEM) techniques. SEM is the surface characterization technique that permits the analysis of a material's external morphology. The electrons in the beam are focused on the surface that interacts with the material, thus generating signals that provide knowledge of surface topography and composition.

One of the major advantages of SEM is the larger field of view ranging from micrometer (μm) to nanometer (nm), which allows a larger amount of the sample to be in focus at one time. The field

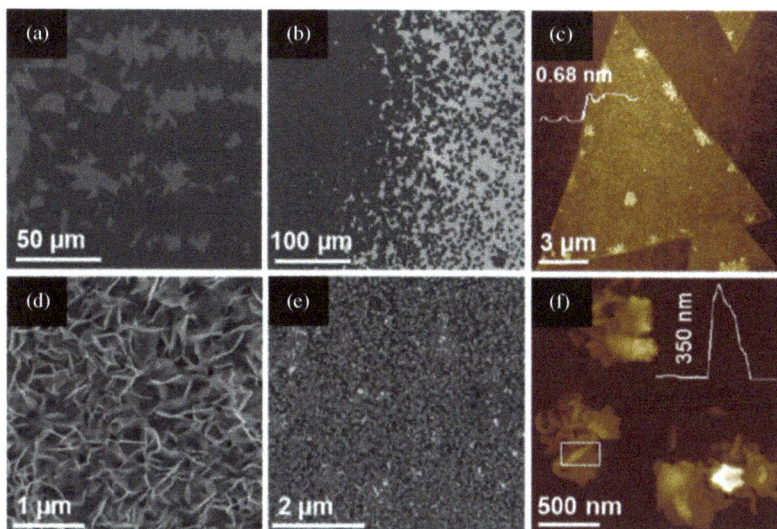

Fig 3.7 (a, b) High- and low-magnification FESEM images of horizontally oriented single-layer MoS$_2$, coalescing to form a continuous film and (c) corresponding AFM image with a height profile of ~0.68 nm. (d, e) High- and low-magnification FESEM images of vertically oriented few-layer MoS$_2$ and (f) corresponding AFM image showing the height of ~350 nm. Reprinted with permission from Kumar and Viswanath [2017]. Copyright (2023) Royal Society of Chemistry.

emission SEM (FESEM) image of CVD-grown monolayer horizontally oriented MoS$_2$ (Fig 3.7(a,b)) and few-layer vertically oriented MoS$_2$ (Fig 3.7(d, e)) reveals their morphology and large area coverage on SiO$_2$/Si substrate [Kumar & Viswanath, 2017]. AFM is used for imaging the smooth surface topography of MoS$_2$ by providing high-resolution 3D topographic images of the material. It is a straightforward technique to find out the thickness of synthesized MoS$_2$ and hence unambiguously identify the exact layer number. Tapping mode is adapted to image the sample and identify the thickness of the layer while contact mode is usually applied for pointing deformation by applying a controlled mechanical force on the sample. Fig 3.7(c) shows the AFM image and the height profile of CVD-grown horizontal MoS$_2$, confirming the thickness of ~0.68 nm, that is, single-layer growth. Fig 3.7(f) shows the AFM image of CVD-grown vertically oriented MoS$_2$ nanostructures with a height profile of ~350 nm, showing the vertical alignment. TEM is another versatile tool applicable for quantitatively identifying the layers, morphology, crystal structure, crystallinity, and defect density, if any. High-resolution TEM (HRTEM) enables a deep analysis with atomic resolution.

3.2.3 *Properties of MoS$_2$*

3.2.3.1 *Electronic Property*

The two-dimensional MoS$_2$ structure displays a unique electrical property due to quantum confinement and transformation of the indirect energy gap of bulk to direct gap for monolayer. The electronic property of MoS$_2$ is also affected by the crystal structure. Zhao and Liu [2018] compared the band structure of different crystal phases (1T, 2H, and 3R) of MoS$_2$ in case of bulk, bilayer, and monolayer. In the 2H-MoS$_2$ phase, the S atom overlaps along the *c*-axis implying strong repulsive force. As a result, it leads to a weaker overlap of wave function leading to the existence of a bandgap, that is, it shows semiconducting behavior. Whereas in the 1T and 3R phases of MoS$_2$, the S atom does not overlap along the *c*-axis. That shows the existence of a strong attractive force, leading to a stronger overlap of the wave function and causing the nonexistence of bandgap, that is, revealing conducting behavior.

Semiconducting behavior can be converted into metallic by phase engineering, such as Li intercalation. Electronic-structure engineering provides the vision for modulating the band structure with layer numbers. The indirect-to-direct crossover results from the local shift of VB hills and CB valleys in the Brillouin zone (Fig 3.8(a–f)). Recently, we theoretically calculated the layer-dependent bandgap of MoS$_2$. The VB maxima and CB minima coincide at K point for monolayer MoS$_2$, showing direct bandgap behavior, featured with a wide bandgap of 1.74 eV. As the number of layers increases, layer interaction increases resulting in the rising of VB hill at the Γ point and down-shifts of CB valley between Γ and K points, thus making it an

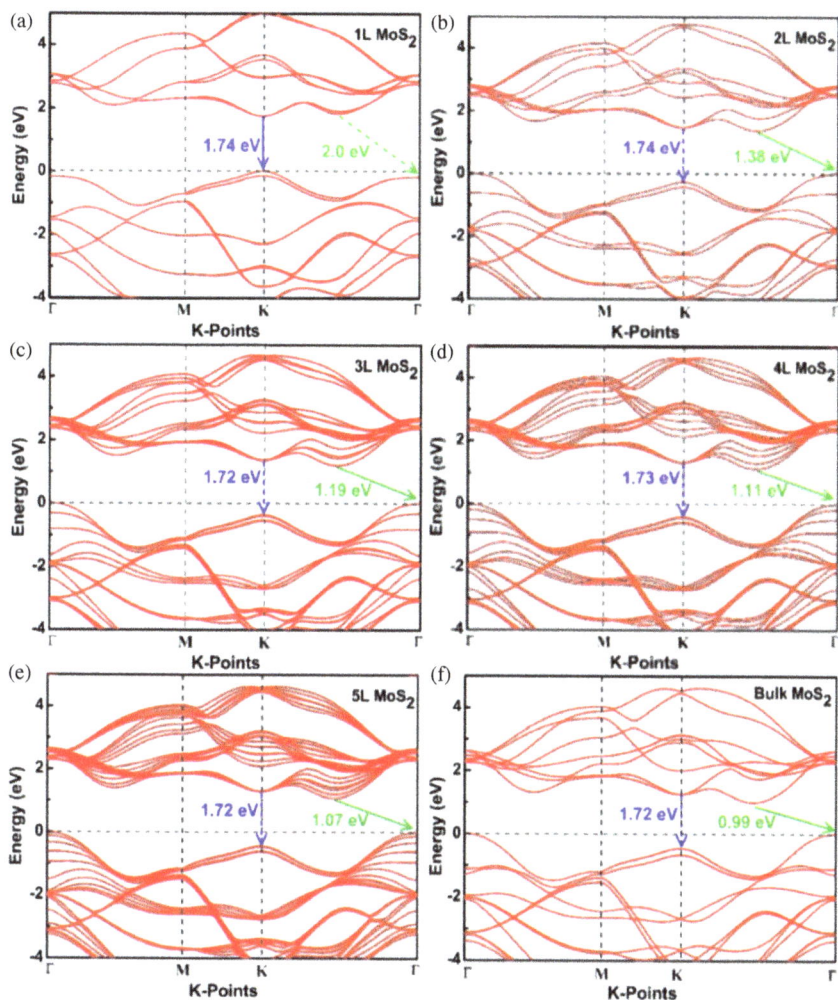

Fig 3.8   Electronic band structures of (a) 1L, (b) 2L, (c) 3L, (d) 4L, (e) 5L, and (f) bulk MoS_2, calculated using DFT. Reprinted with permission from Singh *et al.* [2023]. Copyright (2023) American Chemical Society].

indirect bandgap semiconductor, with 0.99 eV in case of bulk. As the number of layers decreases, the bandgap increases due to the quantum confinement effect along the *c*-direction of the crystal [Singh *et al.*, 2023]. Also, reports have shown that MoS_2 exhibits good mobility ($\sim$200–700 cm^2 V^{-1} S^{-1}) as well as a high current on/off ratio of $\sim$10^5–10^7 [Santhosh & Madhavan, 2019].

3.2.3.2 *Optical Property*

The optical property of a material defines how the material interacts with light and is directly related to its electrical property. It is crucial in industry and scientific work such as contactless temperature measurement, laser technology, and photovoltaic applications. The penetration depth of a particular wavelength of light in the material before it gets completely absorbed is given by the absorption coefficient. Monolayer and multilayer MoS$_2$ have a high absorption coefficient (α) in the visible spectrum (400–500 nm). Hence a sharp decline after 500 nm suggests that the photodetector made of MoS$_2$ will detect light only below 500 nm. Penetration of light of a certain wavelength through a material is defined by the extinction coefficient (k). In the case of monolayer MoS$_2$, k peaks at $\sim$450 nm, that is, most monolayer MoS$_2$ absorbs the maximum of this particular wavelength of light, and for $\lambda > 500$ nm, k has a lesser value showing the transparency in that range. At $\sim$400 nm, multilayer MoS$_2$ shows a larger peak than a monolayer, signifying that at this value of λ, a multilayer absorbs more light than a monolayer. The relation between k and absorption coefficient (α) is given by [Ullah *et al.*, 2018],

$$k = \frac{\alpha \lambda}{4\Pi} \tag{3.1}$$

where λ is the wavelength of light. Another unique feature of MoS$_2$ is the formation of negative trion on illumination. It is a quasiparticle comprising an exciton and an electron. This phenomenon is limited for monolayer MoS$_2$ because all other direct bandgap semiconductors normally form excitons. Trions are formed as a result of valley- and spin-polarized holes and exhibit a large binding energy (20 meV), produced at room temperature [Mak *et al.*, 2013]. For monolayer MoS$_2$, negative conductivity accounts for the trion formation. It has a short life, a long transport channel, and less mobility. Thus, it plants the base for excitonic and spintronic devices.

3.2.3.3 *Mechanical Property*

The mechanical stability of the sample is another important factor to be considered for application in nanodevices. It involves Young's modulus, bending modulus, yield stress, and buckling. The TMD materials have relatively low mechanical strength compared with 2D carbon materials. Bertolazzi *et al.* [2011] calculated the stiffness along with the breaking strength of single-layer MoS$_2$ using AFM. They calculated the in-plane stiffness of a single-layer MoS$_2$ (~180 ± 60 Nm^{-1}) and Young's modulus (~270 ± 100 GPa). The average breaking strength was found to be ~15 ± 3 Nm^{-1} at an effective strain of 6–11%. The yield stress of ~16.5–15 Nm^{-1} was calculated by nanoindentation experiments. MoS$_2$ had a greater bending modulus compared to graphene due to high resistance in the bending motion and was found to be 9.61 eV [Jiang *et al.*, 2013]. Buckling is calculated by simulating molecular dynamics and analysis of phonons. Euler's buckling theory suggests that the buckling strain is proportional to the ratio of bending modulus to in-plane stiffness. The buckling critical strain depends on the length of the layer and is given by [Jiang, 2014]

$$\varepsilon_c = -\frac{43 \cdot 52}{L^2} \tag{3.2}$$

It is 20 times greater for monolayer MoS$_2$ compared to graphene.

3.2.3.4 *Thermal Property*

Thermal properties of TMD nanostructures are crucial for the design of nanoelectronic and thermoelectric devices. The thermal conductivity of 2D TMDs is moderate as compared to monolayer graphene. It is observed that the thermal conductivity of suspended TMDs is largely quenched when transferred on a substrate due to phonons leaking across the material–substrate interface and interfacial scattering. Recently, we reported the layer-dependent thermal conductivity (k_s) of CVD grown supported one-, three-, and five-layer MoS$_2$

nanostructures as 50 ± 2, 34 ± 2, and 28 ± 2 W m^{-1} K^{-1}, respectively, clearly indicating a decrease in thermal conductivity with increasing layer number [Singh *et al.*, 2023]. Majee *et al.* [2020] showed a higher thermal conductivity of 100 ± 14 W m^{-1} K^{-1} for vertically oriented few-layer MoS$_2$ and suggested minimal substrate effect on these MoS$_2$ to be the reason for high thermal conductivity. The defect site in these materials can also reduce thermal conductivity due to higher phonon-defect scattering.

3.2.4 *Other Transition Metal Disulfides*

In addition to MoS$_2$, other transition metal disulfides can also be synthesized using the above-described methods by replacing the Mo precursor with the respective transition-metal precursor and by changing the reaction temperature. Elías *et al.* [2013] reported the synthesis of large-area mono-, bi-, and few-layer WS$_2$ through a two-step process: WO$_3$ was thermally deposited on SiO$_2$ in physical vapor deposition apparatus, followed by sulfurization at high temperatures (750–950°C). WS$_2$ exhibits semiconducting nature with a direct bandgap of 2.1 eV for monolayer and an indirect bandgap of 1.2 eV for bulk WS$_2$ [Guan *et al.*, 2021]. Their unique bandgap characteristics make them effective for photocatalyst and solar-cell applications. SnS$_2$ is an indirect bandgap semiconductor possessing values of 2.41 and 2.18 eV for monolayer and bulk SnS$_2$, respectively [Gonzalez & Oleynik, 2016]. Zhong *et al.* [2012] reported the synthesis of vertically aligned, SnS$_2$ ultrathin nanosheets on Sn substrate by biomolecule-assisted method. They serve as an ideal material for energy storage, catalyst, photoconduction, and field-emitting devices.

3.3 Transition Metal Diselenides (MSe$_2$)

A family of transition metal diselenides includes compounds such as MoSe$_2$, WSe$_2$ VSe$_2$, and NbSe$_2$. Among all, MoSe$_2$ was extensively investigated for energy generation and storage applications and we will focus majorly on MoSe$_2$ in the following subsections.

3.3.1 *Synthesis of MoSe$_2$*

The synthesis of selenide TMDs such as MoSe$_2$ is facing more challenges than their sulfide counterparts, due to the low reactivity of Se. Similar to MoS$_2$, various methods for fabricating MoSe$_2$ nanostructures were reported and are described in the following subsections.

3.3.1.1 *Mechanical Exfoliation*

Exfoliating MoSe$_2$ film initially requires bulk MoSe$_2$. A thin layer of it is adhered with the help of a scotch tape and then transferred onto the desired substrate. This method is applicable due to weak van der Waal interactions between consecutive Se-Mo-Se layers. Tonndorf *et al.* [2013] mechanically exfoliated monolayer to few-layer flakes of MoSe$_2$, which was assured by Raman spectroscopy and PL emission. Although the high quality of the flake is realized by this route and applicable for proof-of-concept, it holds critical drawback in controlling the thickness and uniformity of the flake along with scalability and mass production for device fabrication. Therefore, to mark the industrial application, it is obligatory to work on a more reliable fabrication technique.

3.3.1.2 *Liquid Phase Exfoliation*

It is one of the popular and successfully applied methods in which bulk MoSe$_2$ is dispersed in the liquid phase to separate the layers bonded by a weak van der Waal interaction with the help of agitations such as shear, ultrasonication, or thermal vibrations, to obtain mnolayer to few-layer MoSe$_2$ nanosheets having large surface area. In short, this process includes immersion, insertion, exfoliation, and stabilization [Shen *et al.*, 2016]. The LPE technique is of two types, namely lithium intercalation and solvent-assisted sonication. In lithium intercalation, lithium-containing compounds are used to intercalate between the layers, followed by treatment with water to detach the layers by producing H$_2$. Zhang *et al.* [2017] synthesized monolayer MoSe$_2$ of thickness 0.65–0.7 nm using a liquid-phase exfoliation technique. MoSe$_2$ powder was dispersed in a mixture of tetrahydrofuran and acetonitrile and was then ultrasonicated for 2 h, followed by centrifugation. This process was repeated and the obtained supernatants were MoSe$_2$ nanosheets. In another exfoliation process, bulk MoSe$_2$ was dispersed in acetone, water, and isopropanol, followed by ultrasonication in an ice-cooled sonicator and centrifugation. As a result, 0.7-nm-thick MoSe$_2$ was obtained [Jha *et al.*, 2019]. The disadvantage of this technique is that obtained small sized flakes cannot be used for the applications where large area is required.

3.3.1.3 *Hydrothermal Method*

This method is very common and widely applied at the lab scale for the facile synthesis of MoSe$_2$ nanostructures. First, the precursors are evenly mixed with a solvent (DI) and heated in a Teflon-lined stainless steel autoclave at high temperature and pressure. Mild growth condition and good crystallization make this method an effective way to fabricate nanomaterials. Fan *et al.* [2014] hydrothermally produced flower-like MoSe$_2$ nanostructures by using Na$_2$MoO$_4$ and Se as precursors in the presence of hydrazine hydrate (N$_2$H$_4$·H$_2$O) and DI. The pH value was maintained as 12 by the addition of NaOH. Reactions in an autoclave were carried out at 180°C for 48 h, with subsequent natural cooling. It was then washed (using ethanol and DI) and dried in a vacuum oven to obtain the MoSe$_2$ nanoflower. Tang *et al.* [2016] opted for the hydrothermal method to synthesize 3D hierarchical MoSe$_2$ microspheres of about 1 μm size by reacting Na$_2$MoO$_4$ and Se powder at 200°C for 48 h. They also proposed the growth mechanism in which the precursors were reduced to Mo^{4+} and Se^{2-} by using NaBH$_4$, which further reacted to form MoSe$_2$. Poor solubility and density of Se lead to poor contact during the reaction. Hence, the organic precursor of Se, that is, selenium cyanoacetic acid sodium (NCSeCH$_2$COONa) was used, Here, Se is in the form of Se^{2-} which can directly react in the presence of ethylene glycol rather than toxic hydrazine hydrate, producing $MoSe_2$ hierarchical microspheres of 1 μm size [Dai *et al.*, 2015].

3.3.1.4 *Chemical Vapor Deposition*

This method has been successfully used to date for the fabrication of various monolayers and few-layer $MoSe_2$ film with scalable sizes, high crystallinity, and excellent electronic properties. In this method, MoO_3 and Se powders are used as Mo and Se precursors, respectively. MoO_3 is kept in the middle of the furnace, while Se powder is kept in the low-temperature zone at the upstream region with flowing carrier gas, that is, a mixture of H_2 and Ar. The cleaned substrate is loaded upside down at the downstream region near MoO_3. For temperatures above the boiling point of Se (684.8°C), the Se vapors chemically react with MoO_3 powder to deposit $MoSe_2$ on the substrate. Xia *et al.* [2014] fabricated a large-area, high-quality, and atomically thin $MoSe_2$ on diverse substrates (SiO_2/Si, mica, and Si) by selenization of MoO_3 at 820°C.

The reaction at 750°C is as follows:

$$H_2 + MoO_3 = Mo + H_2O \qquad (3.3)$$
$$2Se + Mo = MoSe_2 \qquad (3.4)$$

Solid Mo reacting with Se vapor inside the furnace at ~900°C also produces $MoSe_2$. But this process is slow and not scalable. The use of organic precursors can better control the large-area monolayer growth of $MoSe_2$. Etzkorn *et al.* [2005] demonstrated the growth of hollow fullerene such as $MoSe_2$ nanoparticles by MOCVD method using $Mo(CO)_6$ and elemental Se. However, a persistent problem with MOCVD is the unintentional significant carbon contamination of the grown film due to the organic precursors used, which severely affects the device performance [Zhang *et al.*, 2016]. Li *et al.* [2018] reported the variable shape evolution, from hexagon to truncated triangle to final sharp edge triangle, of monolayer CVD grown $MoSe_2$ along with the attributed mechanism depending on the increasing H_2 concentration.

3.3.1.5 *Other Methods*

In addition to the above-mentioned methods, there are also other methods that can be used to grow $MoSe_2$ nanostructures. Ullah *et al.* [2016] presented a pulsed laser deposition (PLD)-assisted selenization process for the growth of continuous monolayer $MoSe_2$. In this method, initially, MoO_3 is formed by controlling laser energy and deposition time before selenization. Another promising method for fabricating highly oriented ultrathin $MoSe_2$ film is molecular-beam epitaxy (MBE). This technique is applied to epitaxially grow thin films on various crystalline substrates. Compared to CVD, this method allows better control over film thickness, dopant concentration, and also the growth of high-quality heterostructures [Roy *et al.*, 2016]. Ohtake and Sakuma [2019] described the MBE synthesis of $MoSe_2$ film on Si (111) substrate. The Si (111) is passivated by a GaSe bilayer that allows the formation of $MoSe_2$ film. Cheng *et al.* [2017] applied a two-step growth process for synthesizing $MoSe_2$ nanoribbons. First, Se/Au template is formed by evaporating Se atoms on Au substrate, followed by codoping of Mo atom and Se to form $MoSe_2$ nanoribbons. Chaturvedi *et al.* [2016] produced $MoSe_2$ by microwave induced-plasma-assisted technique. In this method, the plasma of Se inside the ampoule reacts with the outer tube plasma of Mo. Several other methods were reported but, currently, it is still a challenging task for preparing high-quality $MoSe_2$ film for industrial applications.

3.3.2 *Characterization of MoSe₂*

$MoSe_2$ films obtained by various methods were characterized using XRD, optical and electron microscopy, and Raman and PL spectroscopy.

Fig 3.9(a–e) shows the optical microscopy image of 6L (~5.4 nm), 4L (~3.6 nm), 3L (~2.7 nm), 2L (~1.8 nm), and 1L (~0.9 nm) of $MoSe_2$ film obtained by layer-by-layer plasma-etching process, where L stands for layer [Sha *et al.*, 2017]. Fig 3.9(f) shows Raman

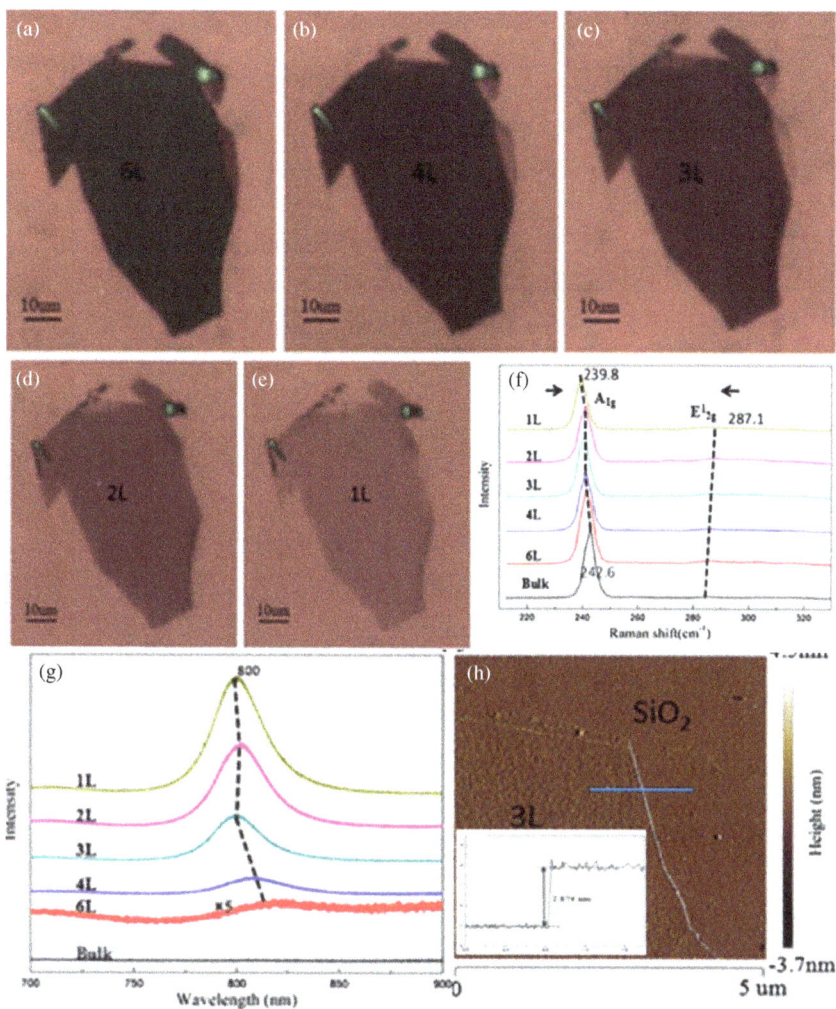

Fig 3.9 (a)–(e) Optical microscopy image, (f) Raman spectra, (g) photolu-minescence (PL) spectra of the etched $MoSe_2$ flakes with different layer numbers, (h) AFM image of 3L etched $MoSe_2$ flake with the height profile (2.74 nm) shown in the inset. Reprinted with permission from Sha *et al.* [2017]. Copyright (2023) Elsevier.

spectra of bulk, 6L, 4L, 3L, 2L, and 1L of the etched $MoSe_2$ film. For both exfoliated and CVD-grown monolayer $MoSe_2$, the A_{1g} mode (out-of-plane) and E_{2g}^1 (in-plane vibration) modes are located at ~240 cm⁻¹ and ~287 cm⁻¹, respectively. Moving from monolayer to bulk, the inter-planar restoring force increases and thus the A_{1g} mode is blue-shifted, while the change in the dielectric screening leads to the red-shift of the E_{2g}^1 mode. The intensity of E_{2g}^1 mode is very weak and is rarely seen. Mostly A_{1g} mode is observed being shifted to the higher wavenumber side with the layer number. The bandgap of $MoSe_2$ is usually confirmed by PL studies. Fig 3.9(g) shows the PL spectra of different layer numbers of etched $MoSe_2$ flake.

Monolayer $MoSe_2$ exhibits the highest intensity peak at 800 nm, attributing to the direct bandgap transition at the K point of Brillouin zone. As the number of layers increases from monolayer to bulk, the bandgap changes to indirect. Also, the peak intensity decreases gradually and the peak position is red-shifted from 800 to 820 nm. In addition, for multilayer $MoSe_2$ film, the intensity of the PL peak is hardly noticeable. The homogeneity and layer number of etched $MoSe_2$ film were also characterized using AFM (Fig 3.9(h)). XRD confirms the crystalline nature of $MoSe_2$. Shelke and Late [2019] synthesized few-layer $MoSe_2$ nanoflowers by hydrothermal method. Fig 3.10(a) shows the XRD pattern of $MoSe_2$, confirming the hexagonal crystal structure. Fig 3.10(b) shows the typical TEM image of prepared $MoSe_2$, revealing its nanoflower morphology with solid interior and core-corona architecture. Lattice fringes and lattice spacing of 0.28 nm for the (100) plane of hexagonal phase of $MoSe_2$ are shown in a high-resolution TEM image (Fig 3.10(c)).

3.3.3 *Properties of MoSe₂*

3.3.3.1 *Optical Properties*

To study the optical absorption spectra, $MoSe_2$ is deposited on a non-conducting glass substrate. Delphine *et al.* [2005] measured the absorption on a 0.7-mm-thick $MoSe_2$ (*t*) at room temperature, with a wavelength of 400–1100 nm. The variation of $(\alpha h\nu)^{1/2}$ with $h\nu$ is

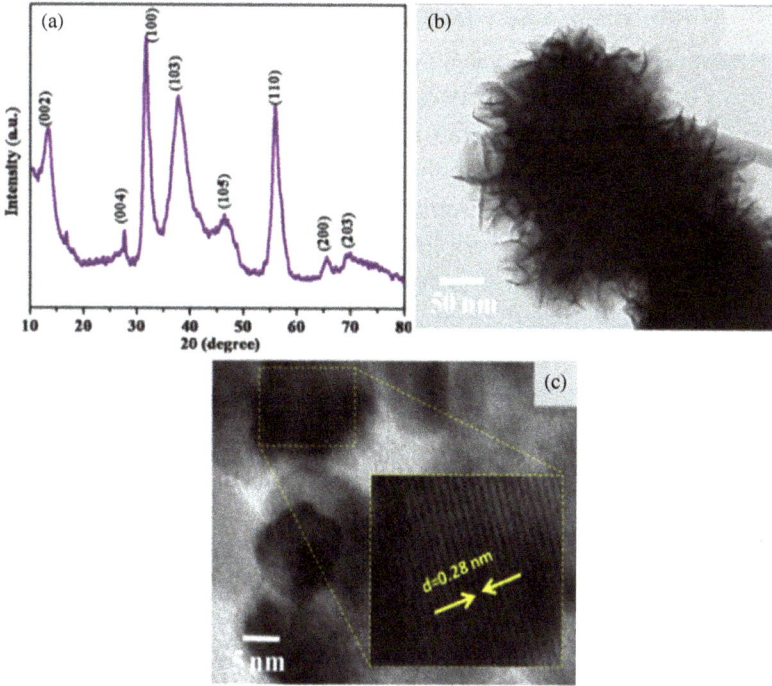

Fig 3.10   (a) XRD pattern, (b) TEM, and (c) HR-TEM image of few-layer $MoSe_2$ nanoflowers. Reprinted with permission from Shelke and Late [2019]. Copyright (2023) Elsevier.

used for calculating the optical bandgap "E_g." Here, α is the absorption coefficient that is a function of radiation energy and composition of the film and can be estimated using the equation $I = I_o\, e^{-\alpha t}$. The linear portion of the plot is extrapolated and bandgap is that point where it cuts the energy axis. It was found to be 1.16 eV for 0.7-mm-thick $MoSe_2$ film.

3.3.3.2 Electrical Properties

The electrical behavior of polycrystalline nanostructures depends on their structural feature and composition. High conductivity shows good crystallinity and thickness of the film. The dark electrical conductivity was measured by choosing a "dc" two-probe method, with the specific conductance at 300 and 500 K being 4.144×10^{-2} and 5.527×10^{-2} Ω^{-1} cm^{-1}, respectively [Hankare *et al.*, 2008]. The specific conductance at room temperature is observed to be $\sim 10^{-2}$ Ω^{-1} cm^{-1}. The electrical conductivity of the film increases with the increase in temperature, signifying the semiconducting behavior but exhibiting variation in the heating and cooling temperature regions. The electrical constants such as activation energy, trapped energy state, and barrier height can be calculated from the electrical conductivity measurements. The activation energy (E_a) is determined with the help of Arrhenius equation:

$$\sigma = \sigma_0 \, e^{-\frac{E_a}{kT}} \tag{3.5}$$

where σ is the conductivity and σ_0 is a parameter that depends on the characteristics of the sample (thickness, structure). The E_a for low- and high-temperature regions are found to be 0.005 and 0.605 eV, respectively.

3.3.4 *Other Transition Metal Diselenides*

Recently, in addition to $MoSe_2$, several other TMDs such as $MoTe_2$, WTe_2, and $TiSe_2$, and their heterostructures have been receiving significant interest in the energy research community. The synthesis procedure of these TMDs is similar to $MoSe_2$. Pazhamalai *et al.* [2016] employed a hydrothermal method to grow $CuSe_2$ nanoneedle on Cu foil and used it as a binder-free electrode for supercapacitors. Zhang *et al.* [2014] described the chemical route for the production of 2D $SnSe_2$ nanodiscs to be used in flexible, all-solid-state supercapacitors. Wang *et al.* [2013] synthesized hierarchical $GeSe_2$ nanostructures by CVD technique that showed a specific capacitance of 300 F g^{-1} at a current density of 1 A g^{-1}. Share *et al.* [2015] displayed WSe_2 as an efficient electrode for sodium-ion batteries for the first time. The exotic properties of 2D-layered TMDs make them applicable for energy storage applications.

3.4 Summary

In summary, we have discussed various synthesis techniques for sulfur and selenium-based TMDs, especially MoS_2 and $MoSe_2$. It is evident that various synthesis methods may result in different 0D, 1D, 2D, and 3D morphologies of these TMDs. These methods have their own merits and demerits in terms of the quality, scalability, and thickness controllability for TMDs, which defines their required potential application. It is observed that the synthesis method and the resultant morphology affect the physical properties of these TMD nanostructures. The excellent electronic properties, fast diffusion kinetics, and high electron conductivity of TMDs make them suitable for energy devices.

References

Ahn, C., Lee, J., Kim, H. U., Bark, H., Jeon, M., Ryu, G. H., Lee, Z., Yeom, G. Y., Kim, K., Jung, J. & Kim, Y. (2015). Low-temperature synthesis of large-scale molybdenum disulfide thin films directly on a plastic substrate using plasma-enhanced chemical vapor deposition. *Adv. Mater.* 27, pp. 5223–5229.

Bertolazzi, S., Brivio, J. & Kis, A. (2011). Stretching and breaking of ultrathin MoS_2. *ACS Nano* 5, pp. 9703–9709.

Campbell, P. M., Perini, C. J., Chiu, J., Gupta, A., Ray, H. S., Chen, H., Wenzel, K., Snyder, E., Wagner, B. K., Ready, J. & Vogel, E. M. (2017). Plasma-assisted synthesis of MoS_2. *2D Mater.* 5, p. 015005.

Chaturvedi, A., Slabon, A., Hu, P., Feng, S., Zhang, K. K., Prabhakar, R. R. & Kloc, C. (2016). Rapid synthesis of transition metal dichalcogenide few-layer thin crystals by the microwave-induced-plasma assisted method. *J. Cryst. Growth* 450, pp. 140–147.

Cheng, F., Xu, H., Xu, W., Zhou, P., Martin, J. & Loh, K. P. (2017). Controlled growth of 1D MoSe2 nanoribbons with spatially modulated edge states. *Nano Lett.* 17, pp. 1116–1120.

Chung, D. Y., Park, S. K., Chung, Y. H., Yu, S. H., Lim, D. H., Jung, N., Ham, H. C., Park, H. Y., Piao, Y., Yoo, S. J. & Sung, Y. E. (2014). Edge-exposed MoS 2 nano-assembled structures as efficient electrocatalysts for hydrogen evolution reaction. *Nanoscale* 6, pp. 2131–2136.

Coleman, J. N., Lotya, M., O'Neill, A., Bergin, S. D., King, P. J., Khan, U., Young, K., Gaucher, A., De, S., Smith, R. J. & Shvets, I. V. (2011). Two-dimensional nanosheets produced by liquid exfoliation of layered materials. *Science* 331, pp. 568–571.

Dai, C., Qing, E., Li, Y., Zhou, Z., Yang, C., Tian, X. & Wang, Y. (2015). Novel MoSe$_2$ hierarchical microspheres for applications in visible-light-driven advanced oxidation processes. *Nanoscale* 7, pp. 19970–19976.

Delphine, S. M., Jayachandran, M. & Sanjeeviraja, C. (2005). Pulsed electrodeposition and characterization of molybdenum diselenide thin film. *Mater. Res. Bull.* 40, pp. 135–147.

Elías, A. L., Perea-López, N., Castro-Beltrán, A., Berkdemir, A., Lv, R., Feng, S., Long, A. D., Hayashi, T., Kim, Y. A., Endo, M., Gutiérrez, H. R., Pradhan, N. R., Balicas, L., Mallouk, T. E., Lo´pez-Urias, F., Terrones, H. & Terrones, M. (2013). Controlled synthesis and transfer of large-area WS2 sheets: From single layer to few layers. *ACS Nano* 7(6), pp. 5235–5242.

Etzkorn, J., Therese, H. A., Rocker, F., Zink, N., Kolb, U. & Tremel, W. (2005). Metal–Organic chemical vapor depostion synthesis of hollow inorganic-fullerene-type MoS$_2$ and MoSe$_2$ nanoparticles. *Adv. Mater.* 17, pp. 2372–2375.

Fan, C., Wei, Z., Yang, S. & Li, J. (2014). Synthesis of MoSe$_2$ flower-like nanostructures and their photo-responsive properties. *RSC Adv.* 4, pp. 775–778.

Gacem, K., Boukhicha, M., Chen, Z. & Shukla, A. (2012). High quality 2D crystals made by anodic bonding: A general technique for layered materials. *Nanotechnol.* 23, p. 505709.

Gonzalez, J. M. & Oleynik, I. I. (2016). Layer-dependent properties of SnS$_2$ and SnSe$_2$ two-dimensional materials. *Phys. Rev. B* 94(12), p. 125443.

Guan, Y., Yao, H., Zhan, H., Wang, H., Zhou, Y. & Kang, J. (2021). Optoelectronic properties and strain regulation of the 2D WS$_2$/ZnO Van der Waals heterostructure. *RSC Adv.* 11(23), pp. 14085–14092.

Hankare, P. P., Patil, A. A., Chate, P. A., Garadkar, K. M., Sathe, D. J., Manikshete, A. H. & Mulla, I. S. (2008). Characterization of MoSe$_2$ thin film deposited at room temperature from solution phase. *J. Cryst. Growth* 311, pp. 15–19.

Jha, R. K., D'Costa, J. V., Sakhuja, N. & Bhat, N. (2019). MoSe$_2$ nanoflakes based chemiresistive sensors for ppb-level hydrogen sulfide gas detection. *Sens. Actuators B.* 297, p. 126687.

Jiang, J. W. (2014). The buckling of single-layer MoS$_2$ under uniaxial compression. *Nanotechnol.* 25, p. 355402.

Jiang, J. W., Qi, Z., Park, H. S. & Rabczuk, T. (2013). Elastic bending modulus of single-layer molybdenum disulfide (MoS$_2$): Finite thickness effect. *Nanotechnol.* 24, p. 435705.

Joensen, P., Frindt, R. F. & Morrison, S. R. (1986). Single-layer MoS$_2$. *Mater. Res. Bull.* 21, pp. 457–461.

Kang, K., Xie, S., Huang, L., Han, Y., Huang, P. Y., Mak, K. F., Kim, C. J., Muller, D. & Park, J. (2015). High-mobility three-atom-thick semiconducting films with wafer-scale homogeneity. *Nature* 520, pp. 656–660.

Kumar, P. & Viswanath, B. (2017). Horizontally and vertically aligned growth of strained MoS$_2$ layers with dissimilar wetting and catalytic behaviors. *CrystEngComm* 19(34), pp. 5068–5078.

Laskar, M. R., Ma, L., Kannappan, S., Sung Park, P., Krishnamoorthy, S., Nath, D. N., Lu, W., Wu, Y. & Rajan, S. (2013). Large area single crystal (0001) oriented MoS$_2$. *Appl. Phys. Lett.* 102, p. 252108.

Li, Q., Walter, E. C., Van der Veer, W. E., Murray, B. J., Newberg, J. T., Bohannan, E. W., Switzer, J. A., Hemminger, J. C. & Penner, R. M. (2005). Molybdenum disulfide nanowires and nanoribbons by electrochemical/chemical synthesis. *J. Phys. Chem. B* 109, pp. 3169–3182.

Li, Y., Wang, F., Tang, D., Wei, J., Li, Y., Xing, Y. & Zhang, K. (2018). Controlled synthesis of highly crystalline CVD-derived monolayer MoSe$_2$ and shape evolution mechanism. *Mater. Lett.* 216, pp. 261–264.

Lin, H., Wang, C., Wu, J., Xu, Z., Huang, Y. & Zhang, C. (2015). Colloidal synthesis of MoS 2 quantum dots: Size-dependent tunable photoluminescence and bioimaging. *New J. Chem.* 39, pp. 8492–8497.

Liu, N., Kim, P., Kim, J. H., Ye, J. H., Kim, S. & Lee, C. J. (2014). Large-area atomically thin nanosheets prepared using electrochemical exfoliation. *ACS Nano* 8, pp. 6902–6910.

Majee, B. P., Bhawna, Singh, A., Prakash, R. & Mishra, A. K. (2020). Large area vertically oriented few-layer MoS$_2$ for efficient thermal conduction and optoelectronic applications. *J. Phys. Chem. Lett.* 11, pp. 1268–1275.

Mak, K. F., He, K., Lee, C., Lee, G. H., Hone, J., Heinz, T. F. & Shan, J. (2013). Tightly bound trions in monolayer MoS_2. *Nat. Mater.* 12, pp. 207–211.

Ohtake, A. & Sakuma, Y. (2019). Heteroepitaxy of $MoSe_2$ on $Si(111)$ substrates: Role of surface passivation. *Appl. Phys. Lett.* 114, p. 053106

Pazhamalai, P., Krishnamoorthy, K. & Kim, S. J. (2016). Hierarchical copper selenide nanoneedles grown on copper foil as a binder free electrode for supercapacitors. *Int. J. Hydrog. Energy* 41, pp. 14830–14835.

Phan, H. D., Kim, Y., Lee, J., Liu, R., Choi, Y., Cho, J. H. & Lee, C. (2017). Ultraclean and direct transfer of a wafer-scale MoS_2 thin film onto a plastic substrate. *Adv. Mater.* 29, p. 1603928.

Ren, X., Pang, L., Zhang, Y., Ren, X., Fan, H. & Liu, S. F.(2015). One-step hydrothermal synthesis of monolayer MoS_2 quantum dots for highly efficient electrocatalytic hydrogen evolution. *J. Mater. Chem.* A 3, pp. 10693–10697.

Roy, A., Movva, H. C. P., Satpati, B., Kim, K., Dey, R., Rai, A., Pramanik, T., Guchhait, S., Tutuc, E. & Banerjee, S. K. (2016). Structural and electrical properties of $MoTe_2$ and $MoSe_2$ grown by molecular beam epitaxy. *ACS Appl. Mater. Interfaces* 8, pp. 7396–7402.

Santhosh, S. & Madhavan, A. A. (2019). A review on the structure, properties and characterization of 2D molybdenum disulfide. *ASET International Conferences IEEE*, pp. 1–5.

Sha, Y., Xiao, S., Zhang, X., Qin, F. & Gu, X. (2017). Layer-by-layer thinning of $MoSe_2$ by soft and reactive plasma etching. *Appl. Surf. Sci.* 411, pp. 182–188.

Share, K., Lewis, J., Oakes, L., Carter, R. E., Cohn, A. P. & Pint, C. L. (2015). Tungsten diselenide (WSe_2) as a high capacity, low overpotential conversion electrode for sodium ion batteries. *RSC Adv.* 5, pp. 101262–101267.

Shelke, N. T. & Late, D. J. (2019). Hydrothermal growth of $MoSe_2$ nanoflowers for photo-and humidity sensor applications. *Sens. Actuator A Phys.* 295, pp. 160–168.

Shen, J., Wu, J., Wang, M., Dong, P., Xu, J., Li, X., Zhang, X., Yuan, J., Wang, X., Ye, M. & Vajtai, R. (2016). Surface tension components based selection of cosolvents for efficient liquid phase exfoliation of 2D materials. *Small* 12, pp. 2741–2749.

Singh, A., Gupta, J. D., Jangra, P. & Mishra, A. K. (2023a). Layered chalcogenides: Evolution from bulk to nano-dimension for renewable
energy perspectives. in: Nanomaterials. (Springer) pp. 177–204.

Singh, A., Majee, B. P., Gupta, J. D. & Mishra, A. K. (2023b). Layer dependence of thermally induced quantum confinement and higher order
phonon scattering for thermal transport in CVD-grown triangular
MoS_2. *J. Phys. Chem. C* 127(7), pp. 3787–3799.

Stacy, A. M. & Hodul, D. T. (1985). Raman spectra of IVB and VIB transition metal disulfides using laser energies near the absorption edges.
J. Phys. Chem. Solids 46, pp. 405–409.

Tanaka, M., Fukutani, H. & Kuwabara, G. (1978). Excitons in VI B transition metal dichalcogenides. *J. Phys. Soc. Japan* 45, pp. 1899–1904.

Tang, H., Huang, H., Wang, X., Wu, K., Tang, G. & Li, C. (2016).
Hydrothermal synthesis of 3D hierarchical flower-like $MoSe_2$ micro-
spheres and their adsorption performances for methyl orange. *Appl.
Surf. Sci* 379, pp. 296–303.

Tang, D. M., Kvashnin, D. G., Najmaei, S., Bando, Y., Kimoto, K.,
Koskinen, P., Ajayan, P. M., Yakobson, B. I., Sorokin, P. B., Lou, J. &
Golberg, D. (2014). Nanomechanical cleavage of molybdenum disul-
phide atomic layers. *Nat. Commun.* 5, pp. 1–8.

Tonndorf, P., Schmidt, R., Böttger, P., Zhang, X., Börner, J., Liebig, A.,
Albrecht, M., Kloc, C., Gordan, O., Zahn, D. R. & De Vasconcellos, S.
M. (2013). Photoluminescence emission and Raman response of mono-
layer MoS_2, $MoSe_2$, and WSe_2. *Opt. Express* 21, pp. 4908–4916.

Ullah, F., Nguyen, T. K., Le, C. T. & Kim, Y. S. (2016). Pulsed laser deposition assisted grown continuous monolayer $MoSe_2$. *CrystEngComm* 18,
pp. 6992–6996.

Ullah, M. S., Yousuf, A. H. B., Es-Sakhi, A. D. & Chowdhury, M. H.
(2018). Analysis of optical and electronic properties of MoS_2 for optoelectronics and FET applications. *AIP Conf. Proc.* 1957, p. 020001

Verble, J. L., Wietling, T. J. & Reed, P. R. (1972). Rigid-layer lattice vibrations and van der waals bonding in hexagonal MoS_2. *Solid State
Commun.* 11, pp. 941–944.

Wang, X., Feng, H., Wu, Y. & Jiao, L. (2013). Controlled synthesis of highly
crystalline MoS_2 flakes by chemical vapor deposition. *J. Am. Chem. Soc.*
135, pp. 5304–5307.

Wang, Q. H., Kalantar-Zadeh, K., Kis, A., Coleman, J. N. & Strano, M. S. (2012). Electronics and optoelectronics of two-dimensional transition metal dichalcogenides. *Nat. Nanotechnol.* 7, pp. 699–712.

Wang, X., Liu, B., Wang, Q., Song, W., Hou, X., Chen, D., Cheng, Y. B. & Shen, G. (2013). Three-dimensional hierarchical $GeSe_2$ nanostructures for high performance flexible all-solid-state supercapacitors. *Adv. Mater.* 25, pp. 1479–1486.

Wu, S., Huang, C., Aivazian, G., Ross, J. S., Cobden, D. H. & Xu, X. (2013). Vapor–solid growth of high optical quality MoS_2 monolayers with near-unity valley polarization. *ACS Nano* 7, pp. 2768–2772.

Xia, J., Huang, X., Liu, L. Z., Wang, M., Wang, L., Huang, B., Zhu, D. D., Li, J. J., Gu, C. Z. & Meng, X. M. (2014). CVD synthesis of large-area, highly crystalline $MoSe_2$ atomic layers on diverse substrates and application to photodetectors. *Nanoscale* 6, pp. 8949–8955.

Yu, Y., Li, C., Liu, Y., Su, L., Zhang, Y. & Cao, L. (2013). Controlled scalable synthesis of uniform, high-quality monolayer and few-layer MoS_2 films. *Sci. Rep.* 3, pp. 1–6.

Zeng, Z., Yin, Z., Huang, X., Li, H., He, Q., Lu, G., Boey, F. & Zhang, H. (2011). Single-layer semiconducting nanosheets: High-yield preparation and device fabrication. *Angew. Chem.* 123, pp. 11289–11293.

Zhang, X., Al Balushi, Z. Y., Zhang, F., Choudhury, T. H., Eichfeld, S. M., Alem, N., Jackson, T. N., Robinson, J. A. & Redwing, J. M. (2016). Influence of carbon in metalorganic chemical vapor deposition of few-layer WSe_2 thin films. *J. Electron. Mater.* 45, pp. 6273–6279.

Zhang, S., Nguyen, T. H., Zhang, W., Park, Y. & Yang, W. (2017). Correlation between lateral size and gas sensing performance of $MoSe_2$ nanosheets. *Appl. Phys. Lett.* 111, p. 161603

Zhang, C., Yin, H., Han, M., Dai, Z., Pang, H., Zheng, Y., Lan, Y. Q., Bao, J. & Zhu, J. (2014). Two-dimensional tin selenide nanostructures for flexible all-solid-state supercapacitors. *ACS Nano* 8, pp. 3761–3770.

Zhao, Z. Y. & Liu, Q. L. (2018). Study of the layer-dependent properties of MoS_2 nanosheets with different crystal structures by DFT calculations. *Catal. Sci. Technol.* 8, pp. 1867–1879.

Zhong, H., Yang, G., Song, H., Liao, Q., Cui, H., Shen, P. & Wang, C. X. (2012). Vertically aligned graphene-like SnS2 ultrathin nanosheet arrays:

Excellent energy storage, catalysis, photoconduction, and field-emitting performances. *J. Phys. Chem. C* 116(16), pp. 9319–9326.

Zhou, M., Zhang, Z., Huang, K., Shi, Z., Xie, R. & Yang, W. (2016). Colloidal preparation and electrocatalytic hydrogen production of MoS_2 and WS_2 nanosheets with controllable lateral sizes and layer numbers. *Nanoscale* 8, pp. 15262–15272.

© 2024 World Scientific Publishing Company
https://doi.org/10.1142/9789811283406_0004

Chapter 4

Carbon-Based and TMDs-Based Materials as Catalyst Support for Fuel Cells

Akshaya S. Nair, Soju Joseph, and R. Imran Jafri

Department of Physics and Electronics
CHRIST (Deemed to be University), Hosur Road,
Bengaluru 560029, India
Email: imran.jafri@christuniversity.in

Abstract

Global energy consumption and environmental pollution caused by the extensive use of fossil fuels have increased the need to look forward to more renewable energy sources. Fuel cell, one of the promising energy conversion devices, has the potential to outsmart the existing devices but has several setbacks to be employed on a larger scale. One of the hindrances is the sluggish oxygen reduction reaction kinetics at the cathode and hence requires electrocatalysts to improve its overall performance. This chapter provides a brief overview of graphene and transition metal dichalcogenides (TMDs)-based composites that have the potential to be used as a catalyst support.

4.1 Introduction

Carbon, the 6th element in the periodic table, exists naturally in different forms (allotropes) as minerals, namely diamond and graphite, whereas other structures of carbon are synthetic. It has four crystalline allotropes, namely carbyne (sp^1), fullerene ("distorted," sp^2), graphite (sp^2), and diamond (sp^3). Among the carbon family, graphene, a monolayer of graphite, has prime importance due to its peculiar properties. It possesses a thickness of a single atom and has a two-dimensional (2D) geometry which appears as a planar sheet having a systematic arrangement of sp^2-bonded carbon atoms, organized in a hexagonal manner and shows remarkable crystalline and electronic properties [Novoselov *et al.*, 2005]. Graphene possesses a high mechanical strength along with unprecedented thermal and electrical conductivities. Its high specific area promotes higher absorption, thus dramatically reducing the size of the devices manufactured. High electrical and thermal conductivities lead to higher sensitivity. In addition, graphene has shown remarkable properties such as (a) quantum hall effect, (b) favorable optical transparency (~97.7%) [Nair *et al.*, 2008], (c) high carrier mobility at room temperature (~10 000 $cm^2V^{-1}s^{-1}$) [Geim & Novoselov, 2007], (d) superior young's modulus (~1 TPa) [Lee *et al.*, 2008], and (e) exceptional thermal conductivity ranging between 3000 and 5000 $Wm^{-1}K^{-1}$ [Balandin *et al.*, 2008]. Graphene sheets rolled into cylinders are termed carbon nanotubes (CNTs). CNTs possess high conductivity, mass transport, and chemical stability and also have better corrosion resistance and higher specific surface area (SA) [Bharti *et al.*, 2017]. However, CNTs are chemically inert and hydrophobic in nature, making them less compatible with metal nanoparticles (NPs). Consequently, a surface modification technique is usually used to improve surface chemistry, allowing better immobilization of metal catalysts. The most common technique is acid treatment, but it may lead to a decrease in the electrical conductivity of the structure [Kaewsai & Hunsom, 2018].

In addition to graphene, TMDs have also attracted a huge interest due to their peculiar behaviors. The general chemical formula for TMDs is MX_2, where M and X denote the transition metals that belong to group 4–10 (e.g., Mo, Zr, Hf, Re, Nb, V, Ti, and W) and chalcogens (S, Te, and Se), respectively. TMDs consist of a transition metal packed in between two chalcogen atomic layers. The introduction of metal oxides to carbon-based mesoporous structures has proven to be a boon for the stability of electrocatalyst, shifting research toward a new branch of electrodes, namely Metal-Organic Frameworks (MOFs). The MOFs, as the name suggests, are combinations of organic and inorganic crystalline pervious substances, consisting of a systematic arrangement of positive metal (M^+) ions which are enclosed within the organic molecules forming a reiterating cage-like array. This array provides MOFs with a large SA ($7000 \ m^2g^{-1}$), homogenous pore structure, tunable porosity, uniformity in the structure at the atomic level, elasticity in its topology, geometry, size, and chemical functionality. This allows the scientists to design unique MOFs with control over the topology of the framework and its porosity ultimately making its way to application in energy storage devices. This flexibility in designing was unlike ever seen before with other electrocatalysts.

4.2 Fuel Cells (FCs)

The FC converts the chemical potential energy of a fuel (i.e., energy stored in molecular bonds) directly into electrical energy. Based on the fuel used and the difference in operating temperatures, FCs are of different types, namely polymer electrolyte membrane FCs, direct methanol FCs, alkaline FCs, phosphoric acid FCs, molten carbonate FCs, solid oxide FCs, and reversible FCs. Among the different types of FCs, polymer electrolyte membrane FCs (PEMFCs) are widely used due to their low operation temperature, scalability, and high energy density. PEMFCs use hydrogen (H_2) as fuel and directly convert the chemical energy of the fuel into electricity with water as a by-product. A schematic diagram of PEMFCs is shown in Fig 4.1. An individual cell requires an anode, a cathode, and a separating polymer-electrolyte membrane such as Nafion. At the anode, the H_2 gas breaks down into H+ and e^- (hydrogen oxidation reaction (HOR)). These protons and electrons reach the cathode through the conducting

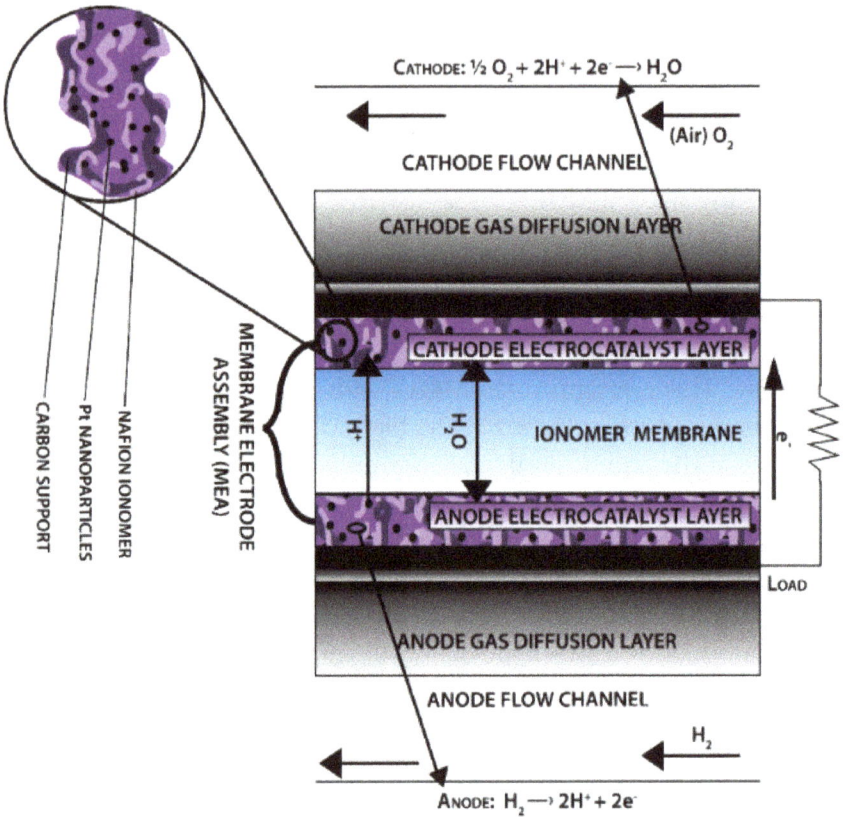

Fig 4.1   Schematic showing the cross section of a typical H_2-O_2 PEMFC.

electrolyte membrane and an external circuit, respectively, and combine with O_2 at the cathode (oxygen reduction reaction (ORR)) resulting in the formation of electricity and water as a by-product, thus completing the electrochemical reaction. Fuels such as methanol, ethanol, and formic acid are also utilized, in addition to hydrogen. Direct methanol FC (DMFC) uses methanol (anode fuel) and combines with oxygen (cathode) to produce electricity and water, thus the reaction happening at anode is referred to as a methanol oxidation reaction (MOR).

The electrocatalyst must be developed so that its catalytic activity increases with respect to cost. Based on the type of FC in consideration, the chemical and physical compositions are determined for the electrocatalyst along with its operating conditions. The *I-V* character-istics are used to study the performance of the electrodes formed by such electrocatalysts. Improving various properties and overcoming the limitations of such catalysts ultimately improves the performance of FC.

4.3 Graphene as Catalyst Support

This section focuses on various graphene-related materials as catalyst support for ORR and MOR that have been exploited in recent years. However, this does not include all the studies done in any way and does not intend to exhaust the subject but rather give an outlook based on the information available. The commercially available FC uses Pt/C as an electrocatalyst and it is noted that, as one moves from bulk to nanoscale, the properties change. It also exclusively focuses on nitrogen-doped graphene, reduced graphene oxide (rGO), function-alized graphene (FG), and catalysts on doped graphene. The effect of the former on the electrochemical surface area (ECSA), the current and power densities, and the impacts of their performances on FCs are reported.

Graphene oxide (GO) is not a very good conductor as it consists of several oxygen-containing functional groups at the edges (carboxyl, carbonyl, quinone groups, etc.,) and on the basal planes (hydroxyl and epoxy groups). Several electroactive species can be used to func-tionalize the functional groups in GO. Therefore, the aim of this study lies in comparing recent and relevant competitors in catalyst support.

4.3.1 *Pt-based Electrocatalysts*

The PtRu/NCNT-GHN also demonstrates a single cell voltage >0.6 V and long-term stability for operation. Ghosh *et al.* [2013] prepared FG via chemical oxidation and thermal exfoliation of natural graphite. The ECSA for Pt/FG was reported to be 57.93 $m^2g^{-1}Pt$. They reported the maximum power density for Pt/FG as 0.455 Wcm^{-2}. Table 4.1 exhibits the results of cyclic voltammetry (CV) up to 1000 cycles.

Table 4.1 ECSA, forward peak current durability test of Pt loaded on three different materials.

	1st Scan			1000th Scan		
	Pt/C	Pt/G	Pt/FG	Pt/C	Pt/G	Pt/FG
ECSA ($m^2g^{-1}Pt$)	40	53.96	57.93	32.22	47.3	53.33
% Reduction in ECSA compared to the initial scan of each catalyst				24.14	14.08	8.62
Forward peak current density ($mAcm^{-2}$)	40.5	43.96	45.78	34.48	39.12	42.06
% Reduction in ECSA compared to the initial scan of each catalyst				17.45	12.37	8.84

Source: [Ghosh *et al.*, 2013]

Jafri *et al.* [2015] synthesized nitrogen-doped graphene samples at different temperatures (180°C and 800°C) via hydrothermal route (NG180) and solid-state method (NGA800) using ammonia and melamine respectively, followed by loading Pt on both. The ECSAs were reported to be 36 and 32 $m^2 g^{-1}$ for Pt/NG180 and Pt/NGA800, respectively. The addition of multiwalled carbon nanotubes (MWCNTs) into the samples showed that the methanol oxidation current densities were 318 and 278 mA $mg^{-1}Pt$ for Pt/NG180 and Pt/NGA800, respectively. Similarly, the ratios of the forward peak current density to the backward peak current density (I_F/I_R) were reported to be 0.97 and 0.89 for Pt/NG180 and Pt/NGA800, respectively, which were considerably higher than Pt/C (0.70) and only a slight decrease in I_F/I_R ratio of Pt/NGA800 was noticed (0.89–0.78).

The maximum power densities were reported as 366 and 328 mW cm^{-2} at 70°C for Pt/NG180 and Pt/NGA800, respectively, and the addition of MWCNT resulted in the increase in maximum power density to (704 $mWcm^{-2}$ (Pt/NG180) and 650 $mWcm^{-2}$ (Pt/NGA800)). Thus, doping with N can lead to significant alterations in the electrocatalytic activity of graphene, and with the addition of MWCNT imprved the utilization of catalyst as active sites became more accessible. Puthusseri and Ramaprabhu [2016] used a polyol reduction technique to prepare nitrogen-doped carbon using Pt as a

Table 4.2 Electrocatalytic performance of Pt/NC at various temperatures with different backpressures.

Backpressure (psi) (cm^2)	Maximum Power Density (mWcm^{-2})	
	At 40°C	At 50°C
0	251	313
5	286	349
10	303	372
15	360	382

Source: [Puthusseri & Ramaprabhu, 2016]

catalyst. They reported that the ECSA for Pt/NC was 24 m^{2}g^{-1}. The sample showed a current density above 367 mAcm^{-2} and the results obtained are shown in Table 4.2. Zhao *et al.* [2016] synthesized 3D hierarchical N-doped graphene (3D-NG) by adopting a solution dip-coating technique, using commercially obtainable sponges. Thus, ECSA was 52.2 m^{2}g^{-1}, which was higher than Pt/3D-G and Pt/G. A superior mass activity for MOR (551.5 mAmg^{-1}) was acquired due to its 3D structure, which was higher than any of its predecessors.

CV studies showed that forward peak current density was retained at 70% after 1000 cycles. Fig 4.2 suggests that Pt/3D-NG (a) is a suitable catalyst for ORR (better diffusion-limiting current density) and (b) has a superior positive onset potential. Bharti *et al.* [2017] developed a superior Pt/CNT catalyst for ORR via microwave-assisted facile method. They followed two distinct heating pathways: (1) conventional heating (Pt/CNT-CH), and (2) microwave heating (Pt/CNT-MWH). Pt/CNT-CH and Pt/CNT-MWH catalysts possessed an ECSA of 48 m^2 g^{-1} and 70 m^2 g^{-1} respectively. Pt/CNT-MWH with higher ECSA is due to an immediate effect of reduced catalyst size achieved via irradiation of microwaves. Higher current densities of 3.9 mA cm^{-2} and 1.3 mA cm^{-2} were shown for Pt/CNT-MWH and Pt/CNT-CH, respectively, at 0.9 V. In addition, the superior reduction current (26%) for Pt/CNT-MWH after 9000 s of operation exhibited higher activity and good stability. The peak power

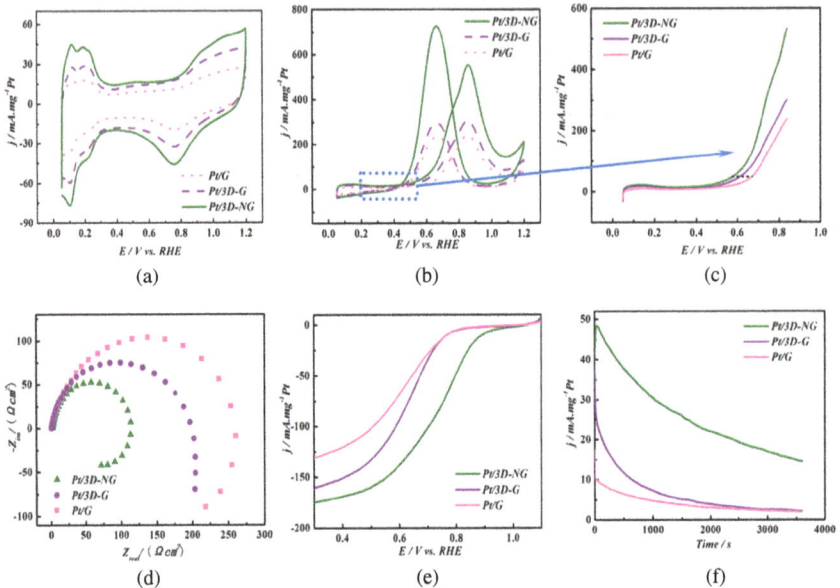

Fig 4.2 (a) CV curves: Pt/3D-NG, Pt/3D-G, and Pt/G catalysts attained at room temperature, (b) mass activity, (c) corresponding potential for a specified oxidation current density, (d) electrochemical impedance spectroscopy (EIS), (e) onset potential and half-wave potential in acidic medium, and (f) current density Reprinted with permission from Zhao *et al.* [2016]. Copyright (2023) American Chemical Society.

densities for Pt/CNT-CH and Pt/CNT-MWH catalysts were also reported as 270 mW cm^{-2} and 365 mW cm^{-2}, respectively.

Kaewsai and Hunsom [2018] studied the ORR activity and stability of Pt and Pt-M (where M can be Ni, Cr, Co, Pd) supported on polyaniline/CNT (PtM/PANI-CNT) catalysts. The improved activity was observed for PtPd/PANI-CNT which could be attributed to to the alteration in the geometry and the electronic structure of the Pt catalyst and by changing M in the Pt-M alloy. However, at 0.6 V, PtPd/PANI-CNT catalyst displayed a high current density of 402 mA cm^{-2} and a power density of 241 mW cm^{-2}. The power densities were reported to be 356 mW cm^{-2} for Cu$_{45}$Pt$_{55}$/VC, 377 mW cm^{-2} for Cu$_{45}$Pt$_{55}$/rGO, and 480 mW cm^{-2} for Cu$_{45}$Pt$_{55}$/rGO-VC. Pt alloyed with Cu enhanced its activity in addition to the benefit of rGO-VC hybrid support [Yılmaz *et al.*, 2019]. Garapati and Sundara [2019] synthesized highly durable Pt$_3$Sc/PECNTs (partially exfoliated CNTs) via polyol method. This electrocatalyst possessed a very high ECSA of 102.1 m^2 g^{-1} and mass activity of 80.5 mA mg^{-1} and showed a current density of 1442 mA cm^{-2}. This high current density was attributed to the combination of the 2D and 1D framework of PECNTs. Even after 500 potential cycles, the catalyst retained its initial activity and onset potential. Pt$_3$Sc/PECNT cathode catalyst exhibited a power density of 760 mW cm^{-2} at 60°C, thus achieving 39% practical efficiency. The synergistic effect of the robust PECNT morphology along with the Pt$_3$Sc (alloy) NPs attributes to its exceptional electrolytic activity, as the electrocatalyst absorbs oxygen with lower binding energy compared to Pt [Garapati & Sundara, 2019]. It was observed that the addition of CNTs to the catalyst layer changed the pore structure, thus increasing its volume. Thereby, suggesting that the ratio of CNT added directly influences the adhesive strength between the catalyst layer and the gas diffusion layer (GDL), achieving higher performance. However, they suffer from low SA. Recent developments in PECNTs gained attraction due to the combined morphology of 1D CNTs with 2D graphene ribbons, retaining higher conductivity and SA [Garapati & Sundara, 2019]. Ce-MOF was used to fabricate a novel Pt-CeO$_2$/TiN nanotubes (TiN NTs) catalyst via hydrothermal method followed by post-nitriding. The Pt-CeO$_2$/TiN NTs showed an ECSA of 59 m^{2}g^{-1}Pt and showed a peak current density of about 0.67 A mg^{-1}Pt which was three times the commercial Pt/C for MOR [Zhou *et al.*, 2019]. The catalyst also showed great stability by maintaining the ECSA after 2000 cycles. Esfahani *et al.* [2020] synthesized the Pt/NbO/CNT as an electrocatalyst which exhibited high electroactivity reaching up to 57 mAmg^{-1}Pt, much higher than its predecessor Pt/CNT in addition to outstanding durability and excellent performance with Pt loading as low as 0.15 mg cm^{-2}. A maximum power density of 772 mW cm^{-2} was achieved which after the durability test (short-term) showed only a 4% loss. This superior activity was attributed to the anchoring effect between Pt and NbO/CNT supports, which not only improved the durability but also the corrosion resistance of the carbon support. Roudbari *et al.* [2020] observed that the Pt catalysts supported by CNT were functionalized by nitrogen to enhance the ORR performance of the PEMFC. It was seen that the electrode prepared with TETA (triethylenetetramine) as a precursor (Pt/MWCNTs-505 TETA) showed great affinity to ORR. The Pt/MWCNTs-U showed the highest electrolytic performance of 64.48 mWcm^{-2}, while Pt/MWCNTs showed the least performance of 23.92 mWcm^{-2}. These indicate that the Pt/MWCNTs-TETA and PT/MWCNT-MN (malonitrile) electrodes result in the lowest ECSA drop. Thus, it was concluded that the functionalization led to the surface modification of the CNT support ultimately leading to a stronger binding of Pt NPs.

To cut down the cost of FC, replacing Pt with non-precious metals or oxides has turned out to be an effective strategy and has seen a surge in research toward that direction, due to their exuberant properties. The following section (4.3.2) covers some of the electrocatalysts that have shown considerable competence with widely used Pt/C.

4.3.2 *Non-Pt Based Electrocatalyst*

Wang *et al.* [2011] demonstrated poly(diallyldimethylammonium chloride) (PDDA) FG as an ORR catalyst. They reported that the ORR current density initially decreased with time. However, it decreased much slower in the case of PDDA-graphene electrode compared with the commercial electrode until finally leveling off (after 17,000 s). This catalyst showed better fuel selectivity and electrocatalytic activity toward ORR.

Fig 4.3 indicates that the onset potential and the ORR reduction peak upon functionalization/adsorption of graphene with PDDA shifted positively. However, the electrocatalytic activity toward ORR with respect to the onset potential and current density for PDDA-graphene was found to be lesser than the commercial Pt/C catalyst. Wu *et al.* [2015] studied the synthesis of graphene via CVD, followed by post-doping it with solid nitrogen precursor of graphitic carbon nitride (g-C$_3$N$_4$) at 700°C, 800°C, and 900°C.

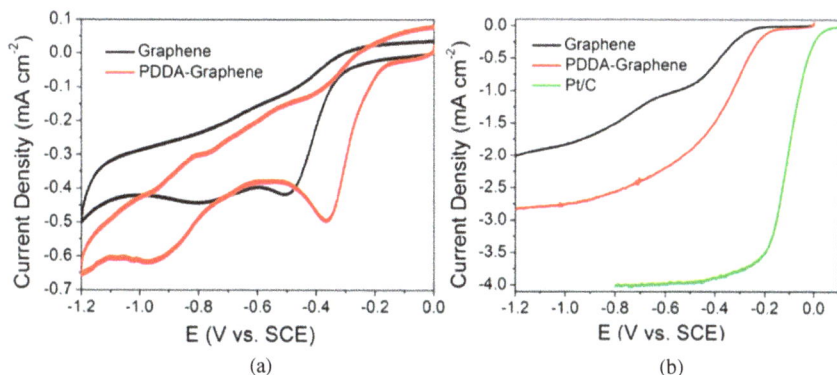

Fig 4.3 (a) CV curves of ORR on graphene and PDDA-graphene electrodes and (b) LSV curves for ORR on graphene, PDDA-graphene, and Pt/C electrodes. [Reprinted with permission from Ref. [Wang *et al.*, 2011]. Copyright (2023) American Chemical Society].

The synthesized nitrogen-doped graphene (NG) constituted higher content of pyridinic-N configuration. The onset potentials for pristine graphene, NG700, NG800, and NG900 were 0.72, 0.92, 0.97, and 0.80 V respectively and it is noteworthy that NG800 has a comparable onset potential as that of Pt/C (0.99 V). Moreover, the author reported that the NG700 and NG800 proceed via a 4e$^-$ process whereas a 2e$^-$ pathway was noted for NG900. The electrocatalytic activity of NG at elevated temperatures is depicted in Fig 4.4. Thus, this study suggests that ORR tends to proceed at lower overpotentials when the pyridinic-N configuration is dominant. An easy and effective method for synthesis of Pd on Au supra-nanostructure (Au@Pd-SprNS)-decked GO nanosheets was adopted by growing Pd crystals on the surface of Au nanoparticles and was modified by a mixture of cetyltrimethylammonium bromide and 5-bromosalicylic acid. A forward peak current density for Au@Pd-SprNSs was reported as 2.24 mA cm^{-2} and 1.89 mA cm^{-2} (with and without GO respectively) and 1.93 mA cm^{-2} for Au@Pd-SprNSs/CA [Tao *et al.*, 2016].

At the 200th cycle, there was a 30% loss reported in the activity of the Au@Pd-SprNSs/GO catalyst. Chronoamperometric (CA) measurements showed that after 1000 s, the current density of Au@

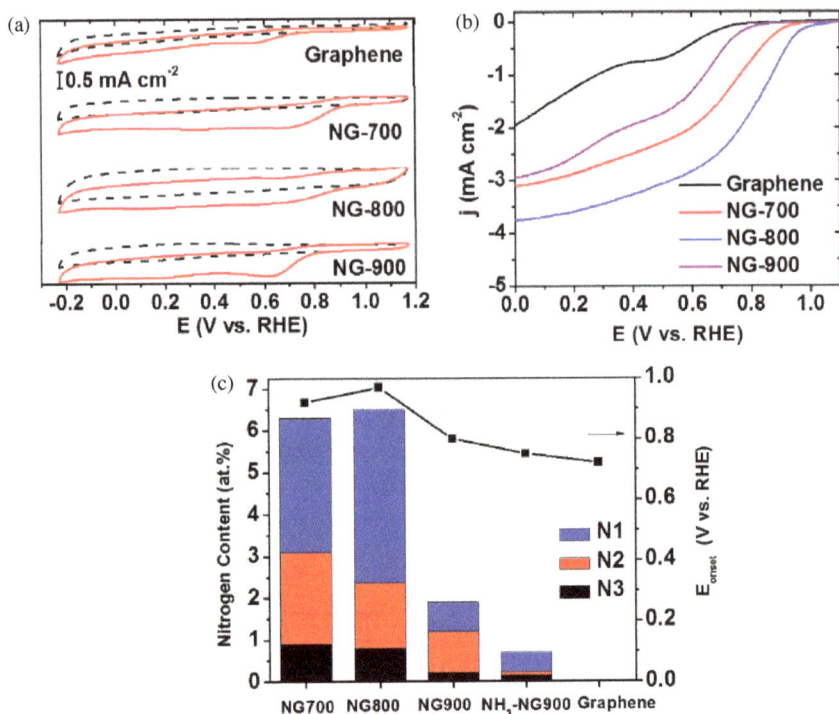

Fig 4.4   Electrocatalytic activity toward ORR of G and NG. (a) CV, (b) RDE polarization curves, and (c) the reliance of onset potential on the N functionality deposition. Reprinted with permission from Wu *et al.* [2015]. Copyright (2023) American Chemical Society.

PdSprNSs/GO was much higher than Pd/C, thus, showing that the catalyst possesses high electrocatalytic durability. Fig 4.5 (a) shows Au@Pd-SprNSs/GO catalyst has higher catalytic activity for ethanol oxidation compared with commercial Pd/C catalyst; (b) the current density of ethanol oxidation on Au@Pd-SprNSs/GO is higher throughout the entire time span with much slower decay in the current density; (c) the decrease in I_F of Au@Pd-SprNSs/GO is lesser than Pd/C catalyst; and (d) the activity loss was ~60% in Pd/C (200th cycle), whereas only 30% in Au@Pd-SprNSs/GO catalyst from the initial. Co-doped Ceria (Co-CeO$_2$) was synthesized via a simple one-step hydrothermal method. Co-CeO$_2$ showed an increase in fuel

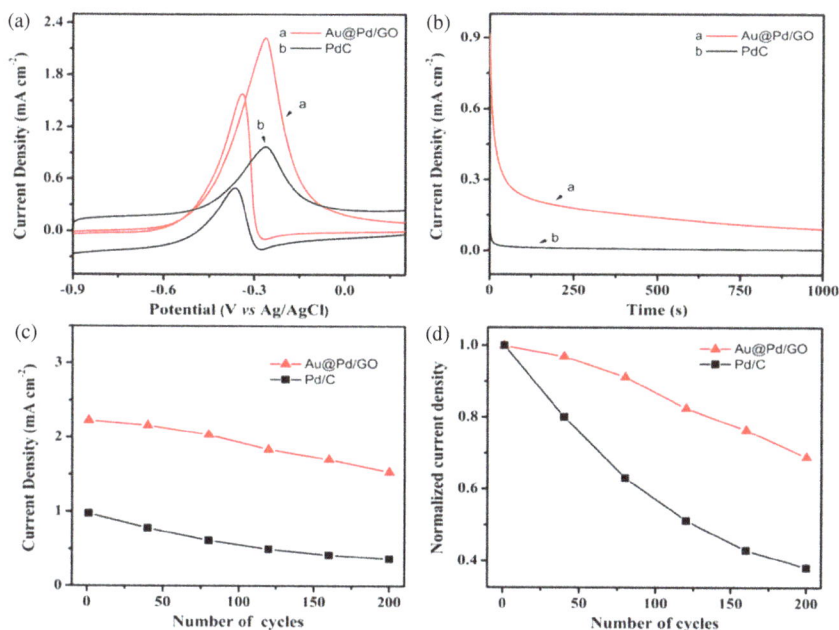

Fig 4.5 (a) CV, (b) chronoamperometry curves of different catalysts and (c, d) long-term stability tests in 1.0 M NaOH + 0.5 M ethanol solution at 50 mV s^{-1} scan rate. Reprinted with permission from Tao *et al.* [2016]. Copyright (2023) American Chemical Society.

selectivity toward ORR and high methanol tolerance even after 10,000 s. Moreover, only a 12% reduction in the initial value of current was found in the Co-CeO$_2$/rGO nanocomposite [Parwaiz *et al.*, 2017].

Fig 4.6 shows a better fuel selectivity toward ORR in a 3% Co-CeO$_2$/rGO nanocomposite compared to commercial catalyst Pt/C [Parwaiz *et al.*, 2017]. Noor *et al.* [2019] synthesized a copper benzenetricarboxylic acid MOF (Cu-BTC MOF) through a facile hydrothermal method and the effect of the addition of GO was studied. The current density of the unhybridized catalyst was 48.55 mA cm^{-2}, which increased to 120 mA cm^{-2} on the addition of GO, indicating that the presence of GO increased the conductivity and thus the entire MOR reaction required less activation energy. Stability studies showed that after 3600 s, the current density of GO

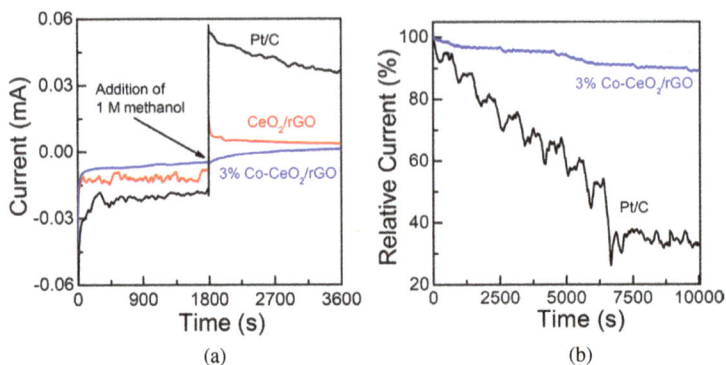

Fig 4.6 (a) CA curve of Pt/C, CeO$_2$/rGO, and 3% Co-CeO$_2$/rGO nano-composites at –0.3 V on adding methanol and (b) CA curve of Pt/C and 3% Co-CeO$_2$/rGO nanocomposites at –0.3 V in 1 M KOH. Reprinted with permission from Parwaiz *et al.* [2017]. Copyright (2023) American Chemical Society.

hybridized catalyst decreased to only 106.42 mA cm^{-2}. Xu *et al.* [2019] prepared a novel Fe-N-C catalyst (ZIF/MIL-10-900) which showed an enhancement in DMFC performance with 3 M methanol solution opening up a pathway to faster kinetics under higher concentrations. It showed a threefold increase in the peak power density compared to Pt/C of about 83 mW cm^{-2}. This suggests that such catalysts are a breakthrough in this field, reducing both the fuel weight and the volume for better efficiency. The stability tests showed that even after 4000 s of testing, there was only a drop of 4.9% in the current density. A conductive support for methanol oxidation in DMFC by synthesizing Ni catalysts doped in zeolitic imidazolate framework (ZIF-67) using a direct non-template technique [Asadi *et al.*, 2019]. The EIS data showed that the charge transfer rate increases and thus conductivity also increases as the activation process increases. This electrode also showed an increase in the cathodic and anodic current peaks with an increase in the number of cycles. The nanoporosity of ZIF-67/CPE provides a larger surface for better loading of Ni, ultimately creating an abundance of active sites for MOR. A core-shell-structured N-doped CoC$_x$/FeCo@C/RGO hybrid as an electrocatalyst was designed by Fang *et al.* [2019] for

DMFC applications. A peak current density of 0.9233 mA cm^{-2}, a higher durability, and a notable electrocatalytic activity were exhibited by this nitrogen-doped hybrid showing a lower overpotential of 390 mV at 10 mA cm^{-2}. These studies suggest that N-doped CoC$_x$/FeCo@C/rGO is a great substitute as a non-precious catalyst for commercial Pt/C. Another core-shell structured catalyst Cu@Pt/C was derived via hydrothermal synthesis of a Cu-based MOF and a central metal was replaced by Pt. It showed an MOR current density of 624 mAcm^{-2}mg^{-1}Pt, which is four times greater than commercial [Long *et al.*, 2020]. The study exhibited that smaller charge transfer resistance makes the catalyst more ideal for methanol oxidation. MOFs are a boon as they are the leading contenders in replacing non-precious electrocatalysts for large-scale applications. This leads to a reduction in cost, but at the price of efficiency. The research in this field is still in the infancy and needs more attention toward developing stable supports for FCs. A comparative study of different electrocatalysts for ORR and MOR is tabulated in Table 4.3.

4.4 Transition-Metal Dichalcogenides (TMDs) as Catalyst Support

Transition metal chalcogenides (TMDs) are catalytically active and are shown to have higher stability in strong acids and various other catalytic properties originating mainly from their edge sites. Thus much focus has been laid on designing and fabricating TMDs with ample edge sites. MoS$_2$ being one of the most widely probed 2D metal dichalcogenides among CoS$_2$, WS$_2$, MoSe$_2$, and WSe$_2$ — due to their facile production, low cost, extended active edges, and basal planes — makes it suitable for the synthesis of hybrid catalyst [Wang *et al.*, 2014]. Pt dispersed on MoS$_2$/graphene nanocomposites exhibited 5.65 times higher electrocatalytic activity than Pt/C for MOR [Zhai *et al.*, 2015]. A higher CO tolerance for MOR was exhibited by Pt dispersed on porous coral reef-like MoS$_2$/N carbon as an electrocatalyst [Tang *et al.*, 2017]. Other layered metal chalcogenides (selenide family) such as MoSe$_2$ and WSe$_2$ have not undergone much research as catalyst supports, but have a lot of potential to be effective through

Table 4.3 Summary of catalytic performance of various electrocatalysts discussed for ORR and MOR.

Electrocatalyst	ECSA (m^2g^{-1})	Current Density	Power Density	Durability	References
Pt/FG	57.93	45.78 mAcm⁻² (HAD)	0.455 Wcm⁻²	—	[Ghosh et al., 2013]
Pt/NG180	36	318 mAmg⁻¹Pt (MOR)	At 70°C • Pt/NG180, 366 mWcm⁻² • Pt/NG800, 328 mWcm⁻²	—	[Jafri et al., 2015]
Pt/NGA800	32	278 mAmg⁻¹Pt (MOR)	• Pt/NG180+MWCNT, 704 mWcm⁻² • Pt/NG800+MWCNT, 650 mWcm⁻²	—	
Pt/3DNG	52.2	—	—	The CV shows that even after 1000 cycles there is 70% remnant forward peak current whereas the current density is 7 times higher than Pt/3D-G and Pt/G even after 3600 s	[Zhao et al., 2016]
CB/PtRu	71.0	—	—	—	[Yang et al., 2017]
CB/PBI/PtRu	67.8	0.15 A mg⁻¹Pt (MOR)	—		
CB/PBI/PtRu/NC	58.4	—	108 mWcm⁻² (MOR)		

Pt/CNT-CH	48	1.3 mAcm⁻² (ORR) (@ 0.9 V)	270 mWcm⁻²	26% higher reduction current for Pt/CNT-MWH after 9000 s of operation	[Bharti et al., 2017]
Pt/CNT-MWH	70	3.9 mAcm⁻² (ORR) (@ 0.9 V)	365 mWcm⁻²		
Cu₄₅Pt₅₅/VC	73	0.43 Acm⁻² @ 0.6 V	356 mWcm⁻²	—	[Yılmaz et al., 2019]
Cu₄₅Pt₅₅/rGO	56	0.53 Acm⁻² @ 0.6 V	377 mWcm⁻²	—	
Cu₄₅Pt₅₅/rGO-VC	119	0.66 Acm⁻² @ 0.6 V	480 mWcm⁻²	—	
Pt₃Sc/PECNT	102.1	1442 mAcm⁻² (@ 0.5 V)	760 mWcm⁻²	After 500 potential cycles, the catalyst retained its initial activity and onset potential	[Garapati & Sundara 2019]

rational designing and fabrication. Triatom layers (Se–W–Se) of WSe$_2$ are bonded by van der Waals forces. WSe$_2$/C composites were prepared via a simultaneous synthesis method exhibiting a current density of 4.57 mA cm^{-2} in an acidic medium with "n" = 3.84 for ORR. The strong contacts between Pt, WSe$_2$, and porous carbon increase the exposure of active sites, significantly promoting faster adsorption, activation, and reduction of O$_2$ molecules [Pan *et al.*, 2017].

WS$_2$ has a metal atom sandwiched between sulfur atom layers and has strong redox catalysis. This S-M-S structure optimizes the active edge density and leads to high catalytic activity. Moreover, the valence of the S atom is –2 (S^{2-}) in WS$_2$. Vulcan XC-72R altered with WS$_2$ nanocomposite (WS$_2$/C) via a solid reaction process showed the onset potential and limiting current density of 0.78 V and 4.99 mA cm^{-2}, respectively. The number of electrons transferred in the ORR electrode was reported as 3.70. Durability tests also showed that the current density remained at 90% after 20,000 s. The addition of Vulcan XC-72R to modified WS$_2$ improved the electrical conductivity as the synergistic effect led to enhanced electrocatalytic activity and electrochemical stability toward ORR in an alkaline solution [Huang *et al.*, 2018]. The potential of nitrogen (N)- and phosphorus (P)-doped TMDs as high-performing electrocatalysts for ORR was demonstrated. Based on the free energy profiles of 4e$^-$ reduction processes, Singh *et al.* [2018] predicted N-TMDs to be more efficient in catalyzing ORR compared with P-TMDs which have a strong binding of reaction intermediates and prevent ORR. It was seen that doping N or P atom activated the basal plane of the monolayer by creating a nonuniform charge distribution. An innovative non-Pt cathodic catalyst for oxygen reduction reaction was designed, wherein hematite NPs were uniformly encapsulated in MoS$_2$/N doped GNS. This catalyst exhibited a comparable electron transfer number (3.91–3.96) as that of commercial Pt/C. Furthermore, the chronoamperometric current density of α-Fe$_2$O$_3$@MoS$_2$/NGNS had a better retention stability of 96.1% when compared to Pt/C (81.4%) after 30,000 s [Chuong *et al.*, 2018].

The solvothermal method was used to design a hybrid material based on MoS$_2$ and rGO (MoS$_2$-rGO) and was employed as a

Table 4.4   ECSA for Pt/MoS$_2$-rGO and Pt/C.

ECSA (m^{2}g^{-1})	Electrocatalysts	
	Pt/MoS$_2$-rGO	Pt/C
5 cycles	32.0	68.4
10,000 cycles	17.2	29.0
Loss in ECSA (%)	46.2	57.6

Source: [Anwar *et al.*, 2019]

cathode electrocatalyst for PEMFC [Anwar *et al.*, 2019]. Higher electrochemical performance was observed due to their collective contribution from the huge number of exposed edges of MoS$_2$ NPs, graphitic nature, the net-like structure of graphene that prohibited the escape of Pt, and exceptional electronic conductivity of rGO. Table 4.4 gives a brief review on the ECSA of Pt/MoS$_2$-rGO and Pt/C. Thus, exploiting other TMDs with graphene could possibly turn out to be an outstanding alternative to the presently employed Pt/C electrocatalyst.

The research in this field is still in its infancy and further probing in rational designing and optimization is required for it to be used on a larger scale. Ramakrishnan *et al.* [2019] reported the synthesis of MoS$_2$/NrGO via hydrothermal method and Pt NPs were anchored using a wet-reflux strategy.

A schematic representation of the synthesis of Pt@MoS$_2$/NrGO nanohybrid for DAFC applications is shown in Fig 4.7. They reported that even after 500 potential cycles, the loss in catalytic activity was below 20%. Pt@MoS$_2$/NrGO activity for the anodic reaction exhibited a 15 mV decrease in the half-wave potential after 30,000 cycles. Thus, MoS$_2$/NrGO can act as an excellent catalyst for both alcohol oxidation reaction (AOR) and ORR in direct alcohol fuel cell (DAFC) applications. Further, optimization in the catalyst structure could make the proposed catalyst durable for other energy applications. The W-N/C@Co$_9$S$_8$@WS$_2$ materials were developed using a dual-step hydrothermal method, and WS$_2$ nanocrystals were wrapped into an N-doped carbon hollow-frame structure. In addition to an improved

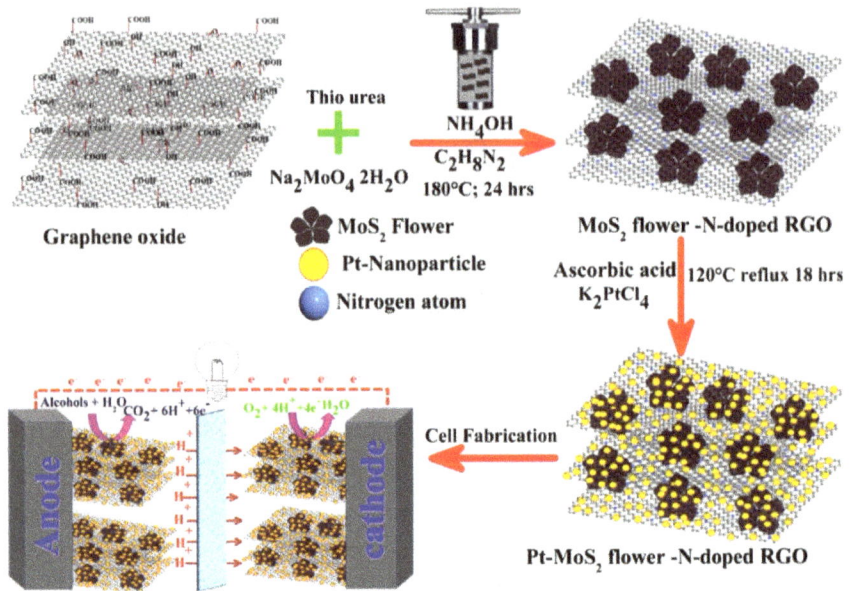

Fig 4.7   Schematic illustration of the synthesis of Pt@MoS$_2$/NrGO nanohybrid for DAFC applications. Reprinted with permission from Ramakrishnan *et al.* [2019]. Copyright (2023) American Chemical Society.

ORR activity, a current density of 10 mA cm^{-2} for oxygen evolution reaction (OER) was obtained at a potential of 1.79 V (for Pt/C it was observed to be 1.91 V) [Liu *et al.*, 2020]. Such an electrocatalyst thus shows promise for ORR and OER.

4.5 Conclusion

The intrinsic activity of the electrocatalysts is essential for the performance of FC, but it is not the only satisfactory parameter. The choice of fuel also alters the performance. This study considers the latest research on basic fuels such as hydrogen and methanol. Heavier fuels such as ethanol, ethylene glycol, ethane, and propane have complex oxidation mechanisms which lead to several intermediate stages ultimately reducing the performance. The primary focus of this study is on various grapheme- and TMD-based catalyst supports and their effects on the performance and stability of FC. Comparative studies based on ECSA, current density, power density, durability, and stability of different electrocatalysts for ORR and MOR are tabulated. It can be seen that moving from the traditional Pt/C commercial electrode to new and novel electrodes by altering the structures of graphene or by tailoring the catalyst itself and using alloys instead of novel metals can turn out to be an advantage, that is, reducing the overall cost and improving the efficiency of FC. It is evident that doping introduces active sites which increases the ECSA and thereby increases the current and the power densities. In terms of durability and stability, the use of alloys instead of noble metals increases durability especially if the alloy is properly fused. Fusion makes it harder for the metals from the alloy to dissolve in the surrounding medium, hence preventing corrosion and increasing both stability and durability. Using CNTs and graphene in turn proved more successful as the combination of 2D and 1D morphologies gives a synergistic effect increasing the overall catalytic performance. In addition, the versatile nature of TMDs has attracted considerable attention toward them due to their higher catalytic activity and electronic structure and has a potential to be an alternative for noble metal catalysts, thus making their way in electrochemical applications. Though a few researches in the field of non-noble catalysts (MOFs and TMDs) are done, they are still in preliminary stages and are yet to achieve the desirable stability and activity required for FCs. But it can be anticipated from the literature that proper synthesis and fabrication techniques can help achieve this target. Thus, the use of tailored graphene with alloy systems is a boon for the FC industry, and further research in this field is required.

References

Anwar, M. T., Yan, X., Asghar, M. R., Husnain, N., Shen, S., Luo, L., Cheng, X., Wei, G. & Zhang, J. (2019). MoS_2-rGO hybrid architecture as durable support for cathode catalyst in proton exchange membrane fuel cells. *Chinese J. Catal.* 40(8), pp. 1160–1167.

Asadi, F., Azizi, S. & Ghasemi, S. (2019). A novel non-precious catalyst containing transition metal in nanoporous cobalt based metal-organic framework (ZIF-67) for electrooxidation of methanol. *J. Electroanal. Chem.* 847, p. 113181.

Balandin, A., Ghosh, S., Bao, W., Calizo, I., Teweldebrhan, D., Miao, F. & Lau, C. (2008). Superior thermal conductivity of single-layer graphene. *Nano Lett.* 8(3), pp. 902–907.

Bharti, A., Cheruvally, G. & Muliankeezhu, S. (2017). Microwave assisted, facile synthesis of Pt/CNT catalyst for proton exchange membrane fuel cell application. *Int. J. Hydrog. Energy* 42(16), pp. 11622–11631.

Chuong, N. D., Thanh, T. D., Kim, N. H. & Lee, J. H. (2018). Hierarchical heterostructures of ultrasmall Fe_2O_3-encapsulated MoS_2/N-graphene as an effective catalyst for oxygen reduction reaction. *ACS Appl. Mater. Inter.* 10(29), pp. 24523–24532.

Esfahani, R. A. M., Vankova, S. K., Easton, E. B., Ebralidze, I. I. & Specchia, S. (2020). A hybrid Pt/NbO/CNTs catalyst with high activity and durability for oxygen reduction reaction in PEMFC. *Renew. Energy* 154, pp. 913–924.

Fang, H., Huang, T., Sun, Y., Kang, B., Liang, D., Yao, S., Yu, J., Dinesh, M., Wu, S., Lee, J. & Mao, S. (2019). Metal-organic framework-derived core-shell-structured nitrogen-doped CoCx/FeCo@C hybrid supported by reduced graphene oxide sheets as high performance bifunctional electrocatalysts for ORR and OER. *J. Catal.* 371, pp. 185–195.

Garapati, M. S. & Sundara, R. (2019). Highly efficient and ORR active platinum-scandium alloy-partially exfoliated carbon nanotubes electrocatalyst for Proton Exchange Membrane Fuel Cell. *Int. J. Hydrog. Energy* 44(21), pp. 10951–10963.

Geim, A. & Novoselov, K. (2007). The rise of graphene. *Nat. Mater.* 6(3), pp. 183–191.

Ghosh, A., Basu, S. & Verma, A. (2013). Graphene and functionalized graphene supported platinum catalyst for PEMFC. *Fuel Cells* 13(3), pp. 355–363.

Huang, H., Zhang, X., Zhang, Y., Huang, B., Cai, J. & Lin, S. (2018). Facile synthesis of laminated porous WS_2/C composite and its electrocatalysis for oxygen reduction reaction. *Int. J. Hydrog. Energy* 43(17), pp. 8290–8297.

Jafri, R., Rajalakshmi, N., Dhathathreyan, K. & Ramaprabhu, S. (2015). Nitrogen doped graphene prepared by hydrothermal and thermal solid state methods as catalyst supports for fuel cell. *Int. J. Hydrog. Energy* 40(12), pp. 4337–4348.

Kaewsai, D. & Hunsom, M. (2018). Comparative study of the ORR activity and stability of Pt and PtM (M= Ni, Co, Cr, Pd) supported on polyaniline/carbon nanotubes in a PEM fuel cell. *Nanomater.* 8(5), p. 299.

Lee, C., Wei, X., Kysar, J. & Hone, J. (2008). Measurement of the elastic properties and intrinsic strength of monolayer graphene. *Science* 321 (5887), pp. 385–388.

Liu, X., Li, X., An, M., karfar, Y., Cao, Z. & Liu, J. (2020). W–N/C@ Co9S8@ WS$_2$-hollow carbon nanocage as multifunctional electrocatalysts for DSSCS, ORR and OER. *Electrochim. Acta* 351, p. 136249.

Long, X., Yin, P., Lei, T., Wang, K. & Zhan, Z. (2020). Methanol electrooxidation on Cu@Pt/C core-shell catalyst derived from Cu-MOF. *Appl. Catal.* 260, p. 118187.

Nair, R., Blake, P., Grigorenko, A., Novoselov, K., Booth, T., Stauber, T., Peres, N. & Geim, A. (2008). Fine structure constant defines visual transparency of graphene. *Science* 320(5881), pp. 1308–1308.

Noor, T., Ammad, M., Zaman, N., Iqbal, N., Yaqoob, L. & Nasir, H. (2019). A highly efficient and stable copper btc metal organic framework derived electrocatalyst for oxidation of methanol in DMFC application. *Catal. Lett.* 149, pp. 3312–3327.

Novoselov, K., Geim, A., Morozov, S., Jiang, D., Katsnelson, M., Grigorieva, I., Dubonos, S. & Firsov, A. (2005). Two-dimensional gas of massless Dirac fermions in graphene. *Nature* 438(7065), pp. 197–200.

Pan, S., Cai, Z., Duan, Y., Yang, L., Tang, B., Jing, B., Dai, Y., Xu, X. & Zou, J. (2017). Tungsten diselenide/porous carbon with sufficient active edge-sites as a co-catalyst/Pt-support favoring excellent tolerance to methanol-crossover for oxygen reduction reaction in acidic medium. *Appl. Catal.* 219, pp. 18–29.

Parwaiz, S., Bhunia, K., Das, A., Khan, M. & Pradhan, D. (2017). Cobalt-doped ceria/reduced graphene oxide nanocomposite as an efficient oxygen reduction reaction catalyst and supercapacitor material. *J. Phys. Chem.* 121(37), pp. 20165–20176.

Puthusseri, D. & Ramaprabhu, S. (2016). Oxygen reduction reaction activity of platinum nanoparticles decorated nitrogen doped carbon in proton exchange membrane fuel cell under real operating conditions. *Int. J. Hydrog. Energy* 41(30), pp. 13163–13170.

Ramakrishnan, S., Karuppannan, M., Vinothkannan, M., Ramachandran, K., Kwon, O. J. & Yoo, D. J. (2019). Ultrafine Pt nanoparticles stabilized by MoS_2/n-doped reduced graphene oxide as durable electrocatalyst for alcohol oxidation and oxygen reduction reactions. *ACS Appl. Mater. Inter.* 11(13), pp. 12504–12515.

Roudbari, M. N., Ojani, R. & Raoof, J. B. (2020). Nitrogen functionalized carbon nanotubes as a support of platinum electrocatalysts for performance improvement of ORR using fuel cell cathodic half-cell. *Renew. Energy* 159, pp. 1015–1028.

Singh, Y., Back, S. & Jung, Y. (2018). Activating transition metal dichalcogenides by substitutional nitrogen-doping for potential ORR electrocatalysts. *Chem. Electro. Chem.* 5(24), pp. 4029–4035.

Tang, B., Lin, Y., Xing, Z., Duan, Y., Pan, S., Dai, Y., Yu, J. & Zou, J. (2017). Porous coral reefs-like MoS_2/nitrogen-doped bio-carbon as an excellent Pt support/co-catalyst with promising catalytic activity and CO-tolerance for methanol oxidation reaction. *Electrochim. Acta* 246, pp. 517–527.

Tao, Y., Dandapat, A., Chen, L., Huang, Y., Sasson, Y., Lin, Z., Zhang, J., Guo, L. & Chen, T. (2016). Pd-on-Au supra-nanostructures decorated graphene oxide: An advanced electrocatalyst for fuel cell application. *Langmuir* 32(34), pp. 8557–8564.

Wang, S., Yu, D., Dai, L., Chang, D. & Baek, J. (2011). Polyelectrolyte-functionalized graphene as metal-free electrocatalysts for oxygen reduction. *ACS Nano* 5(8), pp. 6202–6209.

Wu, J., Ma, L., Yadav, R., Yang, Y., Zhang, X., Vajtai, R., Lou, J. & Ajayan, P. (2015). Nitrogen-doped graphene with pyridinic dominance as a highly active and stable electrocatalyst for oxygen reduction. *ACS Appl. Mater.* 7(27), pp. 14763–14769.

Xu, X., Xia, Z., Zhang, X., Sun, R., Sun, X., Li, H., Wu, C., Wang, J., Wang, S. & Sun, G. (2019). Atomically dispersed Fe-N-C derived from dual metal-organic frameworks as efficient oxygen reduction electrocatalysts in direct methanol fuel cells. *Appl. Catal.* 259, p. 118042.

Yang, Z., Xu, S., Xie, J., Liu, J., Tian, J., Wang, P. & Zou, Z. (2016). Effect of nitrogen-doped PtRu/graphene catalyst on its activity and durability for methanol oxidation. *J. Appl. Electrochem.* 46, pp. 895–900.

Yılmaz, M., Kaplan, B., Gürsel, S. & Metin, Ö. (2019). Binary CuPt alloy nanoparticles assembled on reduced graphene oxide-carbon black hybrid as efficient and cost-effective electrocatalyst for PEMFC. *Int. J. Hydrog. Energy* 44(27), pp. 14184–14192.

Zhai, C., Zhu, M., Bin, D., Ren, F., Wang, C., Yang, P. & Du, Y. (2015). Two dimensional MoS$_2$/graphene composites as promising supports for Pt electrocatalysts towards methanol oxidation. *J. Power Sources* 275, pp. 483–488.

Zhao, L., Sui, X., Li, J., Zhang, J., Zhang, L. & Wang, Z. (2016). 3D hierarchical pt-nitrogen-doped-graphene-carbonized commercially available sponge as a superior electrocatalyst for low-temperature fuel cells. *ACS Appl. Mater.* 8(25), pp. 16026–16034.

Zhou, Q., Pan, Z., Wu, D., Hu, G., Wu, S., Chen, C., Lin, L. & Lin, Y. (2019). Pt-CeO$_2$/TiN NTs derived from metal organic frameworks as high-performance electrocatalyst for methanol electrooxidation. *Int. J. Hydrog. Energy* 44(21), pp. 10646–10652.

Chapter 5

Electrochemical Hydrogen Production Using Carbon-Based and TMDs-Based Nanomaterials As Electrocatalysts

Rohit Kumar Gupta, Prince Kumar Maurya, and
Ashish Kumar Mishra

School of Materials Science and Technology
Indian Institute of Technology (Banaras Hindu University),
Varanasi 221005, India
Email: akmishra.mst@iitbhu.ac.in

Abstract

The increasing global warming, population growth, and energy crisis promote the shift toward highly efficient, nontoxic renewable energy resources. Hydrogen energy is a promising solution to meet energy demand. Among different hydrogen production methods, electrochemical water electrolysis produces complete green hydrogen. Carbon- and transition metal dichalcogenides (TMDs)-based nanomaterials can be used as efficient electrocatalysts for hydrogen production due to their high conductivity, large accessible surface area, flexibility, and good stability in acidic/basic medium. In this chapter, we focus on the effect of morphology, doping, and heterostructure of carbon- and TMD-based electrocatalysts that enhance the performance of electrochemical hydrogen production.

5.1 Introduction

Almost 90% of the energy supply in the world today is produced by carbon-based fossil fuels. Fossil fuel consumption as a source of energy increases global warming, ozone depletion, ocean acidification, resource depletion, and detrimental environmental issues [Rajesh Kumar & Majid, 2020; Li *et al.*, 2023]. These issues have significantly impacted the social and economic development and human, plant, and animal lives across the world. Therefore, developing effective methods for producing renewable energy is essential for an increasing demand for energy [Huld & Amillo, 2015]. Owing to this reason, there was significant progress in converting clean, renewable, and sustainable energy such as wind, solar, hydroelectric, and biomass to supply electric power and carriers of energy storage. To resolve issues related to energy demand, the research community needs to design clean, efficacious, economical, and eco-friendly energy storage and conversion systems.

In this regard, green hydrogen gas (H_2), with a very high calorific value and almost no emissions of dangerous or hazardous gases, was promoted as one of the most promising sustainable and clean energy sources for ecological systems. Natural gas steam reforming is the primary method used in the industrial sector to produce H_2 gas, but this process releases CO_2 into the atmosphere, compromising the environmental benefits of using H_2 gas as a fuel. Therefore, achieving sustainable commercial and industrial hydrogen production is a big challenge. Hydrogen production via electrolysis generates 99.9% pure hydrogen with zero emission of CO_2 compared to other production methods [Crabtree *et al.*, 2004]. This technique has attracted much interest due to its cost-effectiveness, environmental friendliness, high efficiency, and more importantly, zero-percent CO_2 emission. Fig 5.1 depicts the standard schematics for the breakdown of water into hydrogen and oxygen.

Photo-electrocatalytic water splitting is another method to break down water into hydrogen and oxygen using solar radiation. Its major disadvantage is its low conversion efficiency. Therefore, optimal photocatalytic material is required for better production. Noble metal

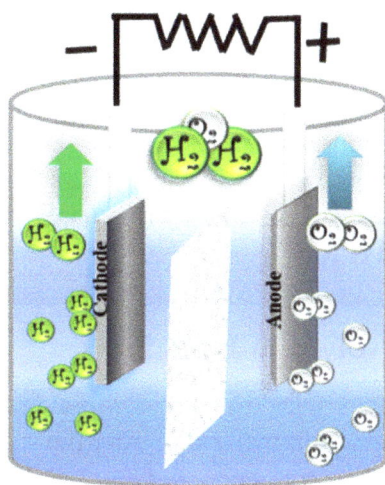

Fig 5.1   Basic schematic for splitting water into hydrogen and oxygen.

(Pt) is generally considered the best electrocatalyst for hydrogen production due to its low overpotential and fast kinetics, but it's high cost and scarcity hinder the use of these catalysts. Therefore, nonmetal catalysts can be utilized, which becomes economically viable but must overcome the problem of high overpotential (η) in the electrochemical process [Kim *et al.*, 2019]. Promising electrocatalysts are required to reduce overpotential values to increase the efficiency and rate of reaction. Developing cost-effective and sustainable electrocatalysts such as metals, metal oxides, metal dichalcogenides, and carbon-based nanostructures as electrocatalysts, is a good choice for better hydrogen production.

The performance of electrochemical hydrogen evolution reaction (HER) depends on the electrocatalytic activity of electrode materials. In particular, good electrocatalytic activity requires the presence of rich active sites and a suitable surface structure. However, pristine carbon nanostructures do not act as promising electrocatalysts for hydrogen production, but they can work as excellent electrocatalysts after surface modification such as functionalization and heterostructure formation. The two-dimensional (2D) transition metal dichalcogenides (TMDs) such as MoS_2, $MoSe_2$, WS_2 and WSe_2

were used as electrocatalysts for hydrogen generation because of their high acidic stability, excellent electrochemical activity, and abundance in nature. These TMDs as electrocatalysts can help to lower the energy required to split the water and enhance hydrogen production. This chapter mainly focuses on fundamentals related to HER and the use of carbon and TMDs nanomaterials as efficient electrocatalysts for HER.

5.2 Reactions Involved in HER Process

Electrochemical water splitting is the most promising clean and green hydrogen production method. Under the standard temperature and pressure, the Gibbs free energy (G^0) of water splitting reaction is 237.2 kJ mol^{-1} [Holmes-Gentle & Hellgardt, 2018] and can be expressed in terms of standard cell potential as follows:

$$\Delta G^0 = -nFE^\circ \tag{5.1}$$

where F is the Faraday constant, n represents the number of transferred electrons per product formed, and E° represents the standard cell potential. Under adiabatic conditions, the enthalpy of water splitting is 286 kJ mol^{-1}. The chemical equation of full water splitting is described by the following reaction:

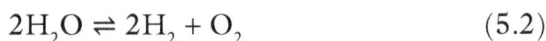

$$2H_2O \rightleftharpoons 2H_2 + O_2 \tag{5.2}$$

Water splitting reaction is divided into two half-cell reactions: cathodic half-cell reaction (reactions 5.3, 5.7) representing proton reduction and anodic half-cell reaction (reactions 5.5, 5.9), showing water oxidation reaction [Moca, 2019]. In acidic electrolyte, the following reactions occur for HER.
    Cathodic reaction:

$$2H^+ + 2e^- \rightleftharpoons H_2, \quad \text{with } E^\circ = 0.00V \tag{5.3}$$

$$E = E^\circ + \frac{0.059}{2}\log[H^+]^2 = E^\circ + 0.0059\log[H^+] = 0 - 0.059\,pH \tag{5.4}$$

Anodic reaction:

$$2H_2O \rightleftharpoons 4H^+ + O_2 + 4e^- \quad \text{with } E° = 1.23 \tag{5.5}$$

$$E = E° + \frac{0.059}{4} \log[H^+]^4 = E° + 0.059 \log[H^+] = 1.23 - 0.059\,pH \tag{5.6}$$

Similar reactions follow for HER in alkaline electrolytes:
Cathodic reaction:

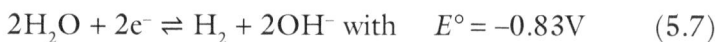

$$2H_2O + 2e^- \rightleftharpoons H_2 + 2OH^- \quad \text{with} \quad E° = -0.83V \tag{5.7}$$

$$E = E° + \frac{0.059}{2} \log[OH^-]^2 = E° + 0.059 \log[OH^-] = -0.83 - 0.059\,pH \tag{5.8}$$

Anodic reaction:

$$4OH^- \rightleftharpoons O_2 + 2H_2O + 4e^- \quad \text{with } E° = 0.40V \tag{5.9}$$

$$E = E° + \frac{0.059}{4} \log[OH^-]^4 = E° + 0.059 \log[OH^-] = 0.40 - 0.059\,pH \tag{5.10}$$

An external voltage is applied across the electrodes to split water into hydrogen and oxygen. The reversible voltage is computed as 1.23 V at 25°C, which is the minimum thermodynamic potential required to electrolyze water. Reactions 5.4 and 5.6 show the influence of pH on the potential in an acidic medium, whereas reactions 5.8 and 5.10 show the influence of pH in an alkaline medium. The Nernst equation reveals that redox potentials shift linearly by −59 mV for each half-reaction. On the other hand, electrochemical processes typically need more energy than thermodynamics estimation.

5.3 Reaction Mechanism for HER

HER is a multistep electrochemical reaction that takes place on the surface of the catalyst, and the mechanism of the hydrogen evolution process follows two-proton-coupled electron transfer mechanisms. The adsorption of proton at the catalyst surface, proton combination, and fast electron transport to the active sites all contribute majorly to this process. The two pathways that govern HER are the Volmer–Heyrovsky and Volmer–Tafel reactions.

5.3.1 HER Reaction in Acidic Medium

The HER mechanism in the acidic medium described by the Volmer reaction is explained by Volmer–Heyrovsky or Volmer–Tafel mechanisms.

Volmer step: Hydronium ion (H_3O+) transfers from a solution to a catalyst surface during the discharge process (Volmer reaction), combining with a cathodic electron to produce H_{ads} on metal sites (reaction 5.11).

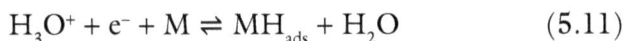

$$H_3O^+ + e^- + M \rightleftharpoons MH_{ads} + H_2O \qquad (5.11)$$

The following two alternative steps can subsequently be taken after the Volmer reaction.

Volmer–Heyrovsky step: H_3O+ in the solution combines with a proton that was adsorbed to the catalyst's surface:

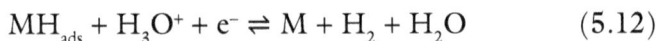

$$MH_{ads} + H_3O^+ + e^- \rightleftharpoons M + H_2 + H_2O \qquad (5.12)$$

The Volmer–Tafel step: Two H_{ads} in the vicinity combine on the catalyst's surface to give molecular hydrogen:

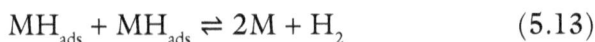

$$MH_{ads} + MH_{ads} \rightleftharpoons 2M + H_2 \qquad (5.13)$$

The active site of the catalyst denoted by M and MH_{ads} represents the adsorbed hydrogen atom at the active site.

5.3.2 HER Reaction in Alkaline Medium

In the alkaline medium, breaking the strong covalent H-O-H bonds takes more energy compared to the weak hydronium ion covalent bond in acidic media. Reactions for HER in the alkaline media are sensitive at the catalyst's surface. The HER mechanism (in the alkaline medium) generally involves two reaction steps, namely Volmer-Heyrovsky or Volmer-Tafel mechanisms.

Volmer reaction: Water adsorption takes place in an alkaline medium. The hydrogen is adsorbed on the catalyst surface due to the coupling of water and e⁻.

$$M + H_2O + e^- \rightleftharpoons MH_{ads} + OH^- \text{ (Volmer step)} \qquad (5.14)$$

Volmer–Heyrovsky reaction: It is an electrochemical desorption in which H_{ads} combines with H_2O molecules.

$$MH_{ads} + H_2O + e^- \rightleftharpoons M + H_2 + OH^- \text{ (Volmer–Heyrovsky step)} \quad (5.15)$$

Volmer–Tafel reaction: Two H_{ads} in the vicinity combine on the surface of the electrocatalyst to give a single molecular hydrogen.

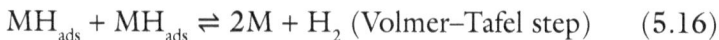

$$MH_{ads} + MH_{ads} \rightleftharpoons 2M + H_2 \text{ (Volmer–Tafel step)} \qquad (5.16)$$

In overall HER reaction, the adsorption and desorption of hydrogen on the catalyst's surface play important roles in either acidic or alkaline medium. The source of hydrogen for HER in acidic and basic electrolytes makes a fundamental difference. Where the active intermediates are generated from protons in alkaline electrolytes, an extra activation mechanism involving the dissociation of water is necessary to generate the H_{ads} on the catalyst surface [Danilovic *et al.*, 2013].

Hydrogen atoms cannot effectively adsorb on the catalyst's surface if the interaction of hydrogen atoms with the catalyst's surface is too weak which delays the HER reaction [Skúlason *et al.*, 2007]. When the interaction is strong between the hydrogen atom and the catalyst's surface, it is difficult to dissociate the hydrogen. Therefore, an optimal interaction is required for the adsorption and desorption of hydrogen. A volcano plot helps to point out the optimal condition for hydrogen adsorption and desorption. As shown in Fig 5.2(a, b), the volcano plot shows the maximum exchange current density of the catalyst where the hydrogen binding Gibbs free energy (ΔG) is

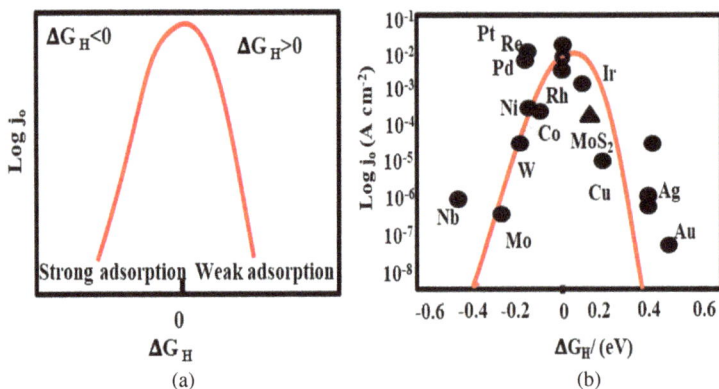

Fig 5.2 (a) Volcano plot predicting catalysts with zero hydrogen binding energy and (b) volcano plot showing probable electrocatalysts for HER. Adapted from Ref. [Parsons, 1958].

minimal or close to zero. The $\Delta G = 0$ represents the volcano peak. Catalyst on the left-hand side of the volcano peak ($\Delta G < 0$) shows a strong interaction of the electrocatalyst surface with H^+. The adsorption of catalyst with H^+ is fast, whereas the desorption is slow. The catalyst on the right-hand side of the volcano peak ($\Delta G > 0$) weakly interacts with hydrogen. The catalyst interaction with active sites is weak. Therefore, an electrocatalyst should have minimum free energy (close to zero) and high exchange current density [Tributsch & Bennett, 1977].

5.4 Performance Evolution Index for Electrocatalyst

The overall electrode activity, overpotential, Tafel slope, turnover frequency, catalytic durability, and faradic efficiency are important parameters that explain the HER catalytic activity of specific electrocatalyst material.

5.4.1 *Overpotential* (η)

One of the most frequent indicators to determine the catalytic activity of an electrocatalyst is the overpotential above and beyond the thermodynamic potential, which drives the reaction at a specific scan rate. Overpotential is the minimal potential required to generate the hydrogen from the electrocatalyst surface. The correlation between the overpotential and the overall potential is given as follows:

$$E_{op} = E_{rev} + iR + \eta \tag{5.17}$$

where E_{op} is the overall potential of electrochemical reactions and iR is the ohmic potential drop due to the resistance of ion flow in the electrolyte. The E_{rev} is the reversible thermodynamical potential and η is an overpotential which is estimated by the polarization curve evaluated using linear sweep voltammetry (LSV) measurements [Zou & Zhang, 2015].

5.4.2 *Tafel Slope and Exchange Current Density*

Tafel slope indicates the rate-determining step of the reaction. It reflects the intrinsic catalytical activity of an electrocatalyst. The Tafel plot indicates a linear relationship between the applied potential and the logarithmic of current density (log j), as given in the following equation:

$$\eta = b \log j + a \tag{5.18}$$

where η represents the overpotential, j denotes the current density (mA cm^{-2}), and b is the Tafel slope (mV dec^{-1}). The high current density suggests that the electrode–electrolyte contact may easily transport electrons with a low activation energy. Evidently, an active electrocatalyst should display a low overpotential and a small Tafel slope for excellent hydrogen production [Benck *et al.*, 2014].

5.4.3 *Electrochemical Impedance Spectroscopy (EIS) Measurements*

EIS is used to understand the electrochemical activity of any material or device. In this method, the charge transfer resistance of the catalyst and factors directly associated with HER activity can be obtained by applying a small potential perturbation and measuring the resulting current response. In the context of the HER, charge transfer resistance represents the hindrance faced by electrons as they participate in the electrochemical reduction of protons to produce hydrogen gas. The higher the charge transfer resistance, the lower the HER performance. EIS data is graphically described by Nyquist and Bode plots. The Nyquist plot consists of three distinct zones: an intercept on the real axis (solution resistance R_s), a semicircle in the high-frequency zone (charge transfer resistance R_{ct}), and a linear component over low-frequency zones (Warburg resistance W) illustrated by the imaginary (Z'') and real (Z') parts of the impedances. The Bode plot is another representation of EIS, which shows the relationship of the magnitude of impedance and phase angle with frequency [Castro *et al.*, 1997].

5.4.4 *Turnover Frequency*

The turnover frequency (TOF) measures the number of catalytic reaction cycles that occur per active site in unit time. It is used to quantify the efficiency of a functional catalytic site at the molecular level. The number of produced hydrogen molecules using TOF is calculated as follows:

$$\text{TOF} = \frac{i}{\frac{n}{N_A}F} = \frac{iN_A}{nF} \tag{5.19}$$

where i is the current density (A cm^{-2}) at a specific overpotential, N_A is the Avogadro constant (6.02×10^{23} mol^{-1}), and F is the Faraday constant (96485 C mol^{-1}). n represents the stoichiometric number of electrons consumed in the electrode reaction (i.e., $n = 2$ for the HER) [Shin *et al.*, 2015].

5.4.5 *Faradic Efficiency (FE)*

FE is a quantitative parameter that reveals the relationship between the moles of products produced at the electrode and the total number of electrons consumed from the external circuit. This is nothing but the ratio of gas evolved experimentally and theoretically at the cathode at a constant current for a given time, as given in the following equation:

$$FE = \frac{\text{Gas evolved (Experimentally)}}{\text{Gas evolved (Theoretically)}} \times 100\% \qquad (5.20)$$

Experimentally produced gas rate estimated under the open circuit potential and on polarization is given in the following equation:

$$FE(\%) = \frac{Q}{n \times F} \times 100\% \qquad (5.21)$$

where F denotes the Faraday's constant ($96{,}485$ C mol^{-1}) and n is the number of electrons produced in the water-splitting reactions. In the catalytic process, $Q(i \times t)$ represents the total electric charge during the reaction [Chen *et al.*, 2021].

5.5 Electrocatalyst for Hydrogen Production

The catalyst's inherent electrochemical behavior and catalytic properties are required to develop more sustainable and low-cost electrocatalysts for hydrogen production. The electrode materials that accelerate reaction rates should have a fast charge transfer rate, maximum active sites, low cost, sustainability, and be abundant in nature. The metals such as Pt-, Pd-, and Ru-based electrocatalysts exhibit high activity toward hydrogen production. However, these metals are quite expensive for large-scale production, and their negative environmental effect hinders their practical application. Hence, much work has gone into finding cost-effective, earth-abundance, and catalytic active alternatives. In this regard, different nonprecious catalysts such as carbon-based materials, transition metal nitrides, transition metal oxides, TMDs, and transition metal phosphides were investigated as electrode material for hydrogen production [Liang & Wu, 2020]. Examples of carbon and TMD nanostructures for hydrogen generation are discussed in the following section.

5.5.1 *Carbon-Based Electrocatalysts for Hydrogen Production*

Carbon nanomaterials, such as graphene, carbon nanotubes (CNTs), and porous carbons, have demonstrated significant potential as electrocatalysts for HER [Hu *et al.*, 2021]. The slow reaction kinetic response of carbon nanomaterials for HER limits the hydrogen production rate. The nature of the electrocatalyst, the pH of the solution, and the concentration of the reactants, all have an impact on the complex kinetics of the HER.

Among carbon nanomaterials, graphene is one of the widely studied electrocatalysts for HER. Graphene has drawn attention in the carbon family due to its distinct 2D structure made up of a single-layer arrangement of carbon atoms in a hexagonally packed lattice. The pristine form of graphene is relatively unreactive (inert nature) and has poor mass transfer capability. Various strategies can be followed to tune its electronic properties and reactivity, rendering its surface catalytically active. Graphene can also serve as a support to disperse other HER active materials, promoting HER kinetics through synergetic coupling effects. Graphene-based HER catalysts possess notable structural advantages such as high electrical conductivity, a large surface area that allows for fast transport of electrolytes to active sites, and chemical stability to withstand harsh electrocatalytic environments [Huang *et al.*, 2019]. Recently developed electrocatalysts based on graphene structures were categorized into several groups for the HER. These categories include heteroatom-doped graphene, introducing defects, graphene/carbon heterostructure, pristine graphene, layered graphene, and other unconventional graphene-based electrocatalysts. These strategies can effectively modulate the electronic structure, chemical reactivity, surface area, and conductivity of graphene, thus enhancing its catalytic efficiency and stability for HER.

Doping introduces heteroatoms (such as nitrogen, boron, sulfur, and phosphorus) into the graphene lattice, which can modulate the electronic properties and create active sites for HER. The presence of nitrogen-containing/doped groups on the surface of carbon nano-structures can improve the catalytic activity for hydrogen evolution. In recent years, N-doped carbon-based materials have attracted a lot of interest because of their excellent qualities and potential uses in a multitude of sectors, including energy storage, catalysis, sensing, and environmental remediation [Ayusheev *et al.*, 2014; Nie *et al.*, 2016]. Integration of nitrogen into carbon materials can modify their electronic structure, which results in improved catalytic activity, increased surface area, enhanced hydrophilicity, and stability. Nitrogen can exist in various forms, such as pyridinic, pyrrolic, graphitic, and quaternary nitrogen, each of which contributes differently to the properties of N-doped carbon materials [Bae *et al.*, 2013; Zhang *et al.*, 2019]. Doping can also enhance the interaction between graphene and metal or metal oxide nanoparticles, which can act as cocatalysts for HER [Duan *et al.*, 2015]. The formation of a nano-structure, which creates defects, edges, pores, or wrinkles on the graphene surface, can increase the exposure of active sites and facilitate charge transfer and mass transport for HER. It can also improve the mechanical flexibility and stability of graphene-based electrocatalysts. In another method, hybridization combines graphene with other materials (such as quantum dots, CNTs, and fullerenes) to form heterostructures or composites, which can synergize the advantages of different components and improve the overall performance of HER. Hybridization can also introduce new functionalities and properties to graphene-based electrocatalysts.

Ma *et al.* [2015] reported the utilization of cobalt phosphite (CoP) nanoparticles equally distributed over sheets of reduced graphene oxide (CoP/RGO) as an electrocatalyst for the HER. In the first stage, GO is dispersed in a mixed solvent with Co^{2+} ions to generate a driving force for the adsorption of Co^{2+} onto GO sheets. When $NH_3 \cdot H_2O$ is added to create an alkaline environment, metal oxide/hydroxide species can form and develop on the functional groups of GO. The resulting Co_3O_4/RGO composite is obtained through solvothermal treatment. In the second stage, to create a uniform CoP/RGO nanocomposite, which has CoP nanoparticles adorned on RGO

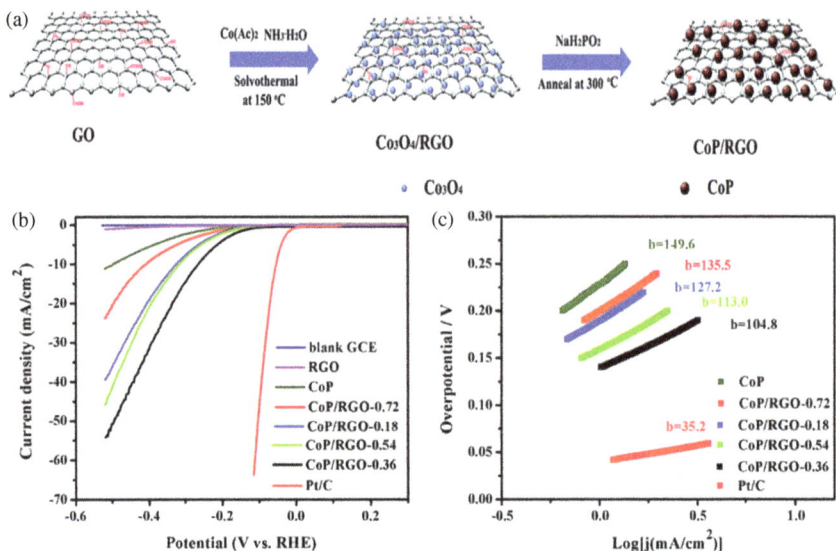

Fig 5.3   (a) Schematic diagram of the synthesis process of CoP/RGO composites, (b) LSV curves, and (c) Tafel plots of prepared materials at a potential sweep rate of 2 mV s^{-1}. [Reprinted with permission from Ref. [Ma *et al.*, 2015]. Copyright (2023) Royal Society of Chemistry].

sheets, the Co_3O_4/RGO composite is annealed at 300°C in the presence of sodium hypophosphite as shown in Fig 5.3(a).

The researchers demonstrated that CoP/RGO-0.36 exhibited excellent electrocatalytic activity of low Tafel slope (104.8 mV dec^{-1}) and exchanged current density of 4×10^{-5} A cm^{-2} at a scan rate of 2 mV s^{-1} compared to CoP alone (Tafel slope and exchange current: 149.6 mV dec^{-1} and 6.3×10^{-7} A cm^{-2}) for the HER, with high stability and durability as shown in Fig 5.3(b, c). The study also highlighted the importance of the interfacial contact between the CoP nanoparticles and reduced graphene oxide sheets in improving the catalytic performance. CNTs are another promising carbon-based electrode material. The catalytic activity of CNTs depends on the structure and surface termination.

Zou *et al.* [2014] discussed the development of cobalt embedded into nitrogen-rich CNTs (Co-NRCNTs) with similar activities to that of platinum for HER. Scanning electron microscopic (SEM) image

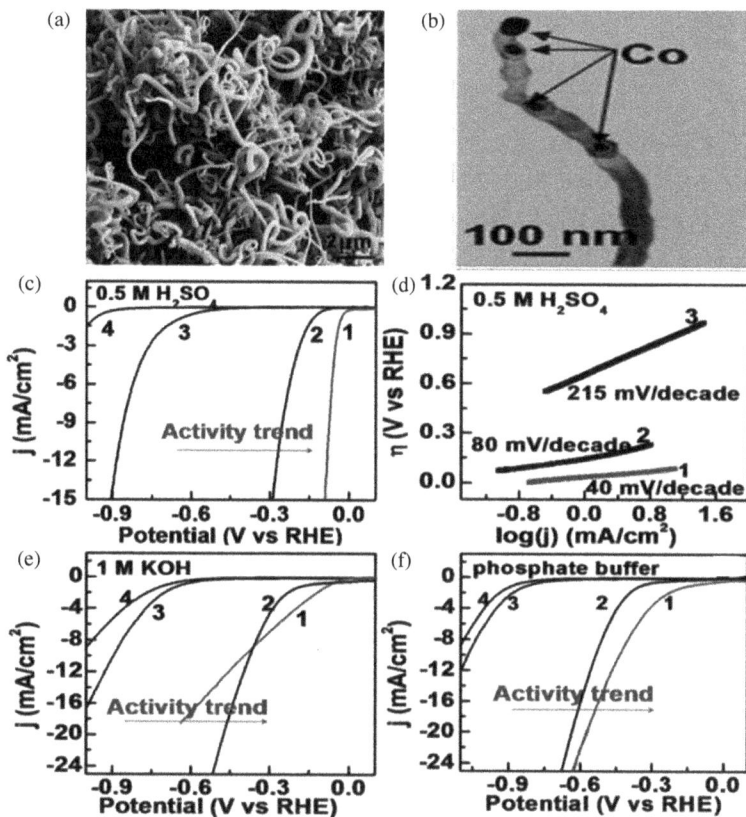

Fig 5.4 (a) SEM image and (b) TEM image; (c) LSV curves and (d) corresponding Tafel slope in acidic media (0.5M H_2SO_4); (e) LSV curves in basic solution (1M KOH); and (f) in neutral media (phosphate buffer) electrolytes of Co-NRCNT. Samples are labeled as 1 (Pt/C), 2 (Co-NRCNTs), 3 (MWCNTs), and 4 (no catalyst/bare). [Reprinted with permission from Ref. [Zou *et al.*, 2014]. Copyright (2023) John Wiley and Sons].

(Fig 5.4(a)) shows that the Co-NRCNTs consist of multiwalled CNTs (MWCNTs) of several meters long and 20–100 nm thick, while the transmission electron microscopic (TEM) image (Fig 5.4(b)) reveals the existence of Co nanoparticles inside the CNTs. Fig 5.4(c–f) describes the LSV curves and Tafel plots for Co-NRCNTs in acidic media (0.5 M H_2SO_4), basic media (1 M KOH), and neutral (phosphate buffer) media. The results show that Co-NRCNTs exhibit a low onset potential of 50 mV and overpotential ($\eta_{10} = 260$ mV) toward HER in acidic media with a cathodic current that rises rapidly as more negative potential is applied, as shown in Fig 5.4(c). The corresponding Tafel slope for Co-NRCNTs shows a low Tafel slope of 69 mV dec^{-1} compared to MWCNTs (Tafel slope = 215 mV dec^{-1}), as shown in Fig 5.4(d). In basic media (1M KOH), Co-NRCNTs show very good electrocatalytic behavior for HER indicating lower onset potential than Pt/C, as shown in Fig 5.4(e). The electrocatalytic activity of Co-NRCNTs in neutral media is shown in Fig 5.4(f). It is relatively lower than in acidic or basic media but is still much better than MWCNTs and is very close to Pt/C catalyst. The materials work well with basic, neutral, or acidic media. The effective catalytic activity of materials is mainly attributable to their structural defects and nitrogen dopants. Mishra *et al.* [2022] discussed MWCNTs as promising electrocatalysts for HER. MWCNTs showed a low onset potential of 162 mV, low overpotential (η_{10}) of 192 mV, and Tafel slope of 105 mV dec^{-1} at a potential scan rate of 2 mV s^{-1}.

5.5.2 *TMDs-based Electrocatalyst for Hydrogen Production*

TMD nanostructures were investigated for use as electrodes due to their exceptional chemical and physical characteristics. They have received great attention for testing their viability as electrode material for energy applications due to abundant active sites, high surface area, morphological dependence, strong ionic conductivity, several valence states, unique electronic structure, and stable semiconducting properties [Singh *et al.*, 2023]. The TMD-based catalysts are widely acknowledged as highly promising alternatives to noble-metal-based catalysts for HER due to their superior catalytic activity and affordable price [Yuan *et al.*, 2020]. Among all TMDs, the fascinating MoS$_2$ nanomaterials with larger specific surface area, extensively exposed active edge sites, excellent hydrophilic nature, and ease of intercalation and modification make them potential candidates for hydrogen production [Lu *et al.*, 2016]. It is well known that the crystal structure of catalysts can reveal information about their active sites. Individual S-Mo-S layers make up the 2D layered structure of MoS$_2$, which has two different types of surfaces: (1) the edge plane, which exposes both Mo and S edge sites, and (2) the basal plane, which only exposes basal S sites. According to several experimental investigations, the catalytic activity of MoS_2 is mostly correlated to their edges rather than the basal planes; however, the edges only make up a small portion of the surface area of MoS_2 layers [Jaramillo *et al.*, 2007]. The electrocatalytic performance of bulk MoS_2 is hindered due to its low intrinsic conductivity and ease of aggregation. The maximum catalytic performance of MoS_2 could be achieved by large exposure of edge sites of unsaturated sulfur atoms compared to its basal plane [Benck *et al.*, 1916].

Wang *et al.* [2013] demonstrated an effective electrochemical catalyst MoS_2 using high-energy ball milling to improve HER activity. Mechanically activated, distorted MoS_2 nanostructures have a maximum number of active edge sites, high active edge planes, and surface area compared to commercial microcrystalline MoS_2. BET surface area measurements make a fair comparison which shows an increase in the surface area from 3.7 to 10.6 $m^2 g^{-1}$ after using a high-energy ball-milling process. The authors measured LSV in 0.5 M H_2SO_4 electrolyte by depositing distorted MoS_2 nanostructure on a glassy carbon electrode. It shows the overpotential and Tafel slope of 200 mV and 104 mV dec^{-1} at 2 mV s^{-1} scan rate with an exchange current density of 1.4×10^{-5} A cm^{-2}. Lower overpotential shows the change of structure with a great amount of active edge site and a smaller Tafel slope indicates the high increment in HER activity as compared to the commercial microcrystalline MoS_2. In case of the durability of MoS_2, there is a slight loss of activity after the 1000 LSV cycles in the acidic environment.

Mishra and Mishra [2020] reported hydrothermally synthesized 2H-phase MoS_2 nanocluster as an alternative to the noble metal as a potent electrocatalyst for hydrogen production. The hydrophilic nature, large electrochemically active surface area, and the maximum number of active edge sites lead to enhancing the kinetics of HER and electrochemical activity. Fig 5.5(a) shows the LSV curve at 2 mV s^{-1} scan rate for hydrothermally synthesized 2H-phase MoS_2 nanocluster. Inset shows the morphology of the hydrothermally

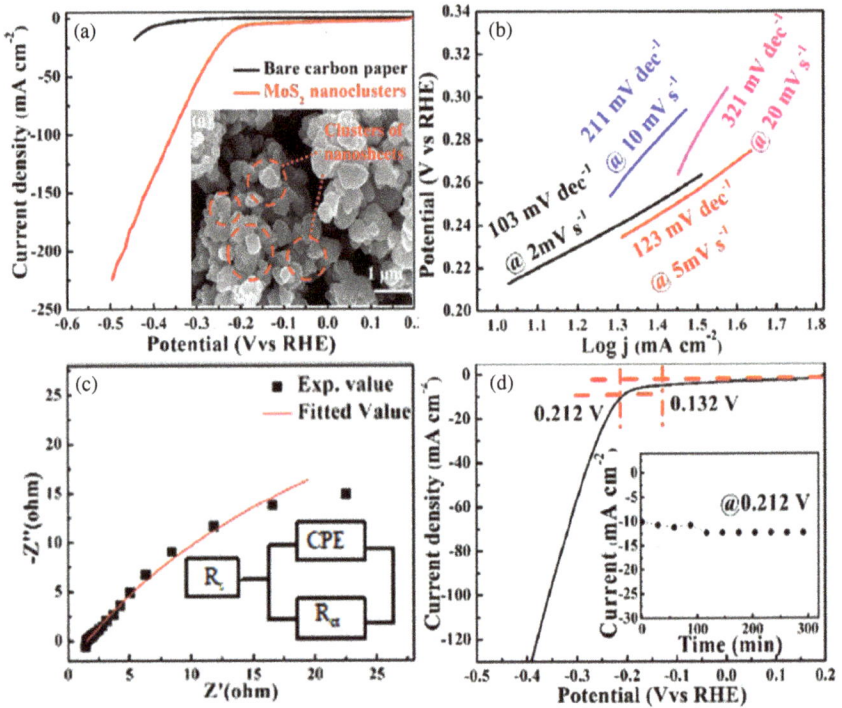

Fig 5.5   (a) LSV curves at 2 mV s^{-1} scan rate and the inset shows the SEM image of MoS$_2$ nanoclusters, (b) Tafel slope at a different scan rate, (c) Nyquist plots, and (d) onset potential and overpotential at 2 mV s^{-1} scan rate with inset showing the stability curve for MoS$_2$ nanoclusters in 0.5 M H$_2$SO$_4$. [Reprinted with permission from Ref. [Mishra & Mishra, 2020]. Copyright (2023) John Wiley and Sons].

synthesized 2H-phase MoS$_2$, which indicates the uniform distribution of nanoclusters. The result shows that the overpotential and Tafel slope decreases as the scan rate decreases due to the maximum number of accessible active sites in the prepared electrocatalyst. Onset potential, overpotential, and Tafel slope are 132 mV, 212 mV, and 103 mV dec^{-1}, respectively. Tafel slope at different scan rates indicates the electrocatalyst's inherent properties and rate determination step for HER, as shown in Fig 5.5(b). In Fig 5.5(c), the charge transfer resistance (R_{ct}) is low, ~83 Ω, which indicates the fast electron-transfer rate for hydrogen production. Fig 5.5(d) shows the onset potential and overpotential at 2 mV s^{-1} scan rate. The inset of Fig 5.5(d) shows the stability plot of the MoS$_2$ nanocluster at a potential scan rate of 20 mV s^{-1} for 300 min, indicating the robustness of electrocatalyst in 0.5 M H$_2$SO$_4$ electrolyte. Cheng *et al.* [2014] reported successfully synthesized ultrathin WS$_2$ nanoflakes using a high-temperature solution-phase method for inexpensive and highly efficient electrocatalysts for hydrogen evolution. The loosely formed ultrathin nanoflakes structure of WS$_2$ shows a low overpotential and Tafel slope of approximately 100 mV and 48 mV dec^{-1} at a potential sweep rate of 2 mV s^{-1}, respectively. The electrocatalytic activity was periodically evaluated and an almost negligible drop in overpotential was observed after 1000 CV cycles for the prepared materials. They also reported monitoring 1000 cyclic CV curves and showed the change in cathodic current density at –0.42 V vs. the saturated calomel electrode (SCE) in an acidic medium (0.5 M H$_2$SO$_4$). The performance of the electrocatalyst is highly stable and does not restack or collapse after a lengthy operation, hence WS$_2$ nanoflakes show excellent electrocatalytic behavior for hydrogen production.

Zhang *et al.* [2021] successfully demonstrated the WSe$_2$ layer with the doping of the Co atom using a hydrothermal route, where the Co atomically displaces the W atom in the WSe$_2$ layers. Element doping is an efficient route for activating the catalytic site and changes in electronic structure. Authors reveal that enhanced hydrogen production for the Co-doped tungsten diselenide (Co-WSe$_2$) nanosheets was observed due to the dynamic structure evolution and electronic state optimization. The doping effect increases and activates the number of active plane sites and improves conductivity, enhancing the electrocatalytic activity and hydrogen production performance. Enhancing and adsorption of H$^+$ ions for hydrogen production takes place due to changes in W-Se and Co-Se bonds. Fig 5.6(a) shows the TEM image of Co-WSe$_2$ nanosheets. The LSV curve (Fig 5.6(b)) shows the overpotential of WSe$_2$ is 345 mV and after the doping of Co atoms, the overpotential of Co-WSe$_2$ is approximately 205 mV at a current density of 10 mA cm^{-2}. Authors also showed that the overpotential increases when the amount of Co atom increases to 6 wt%.

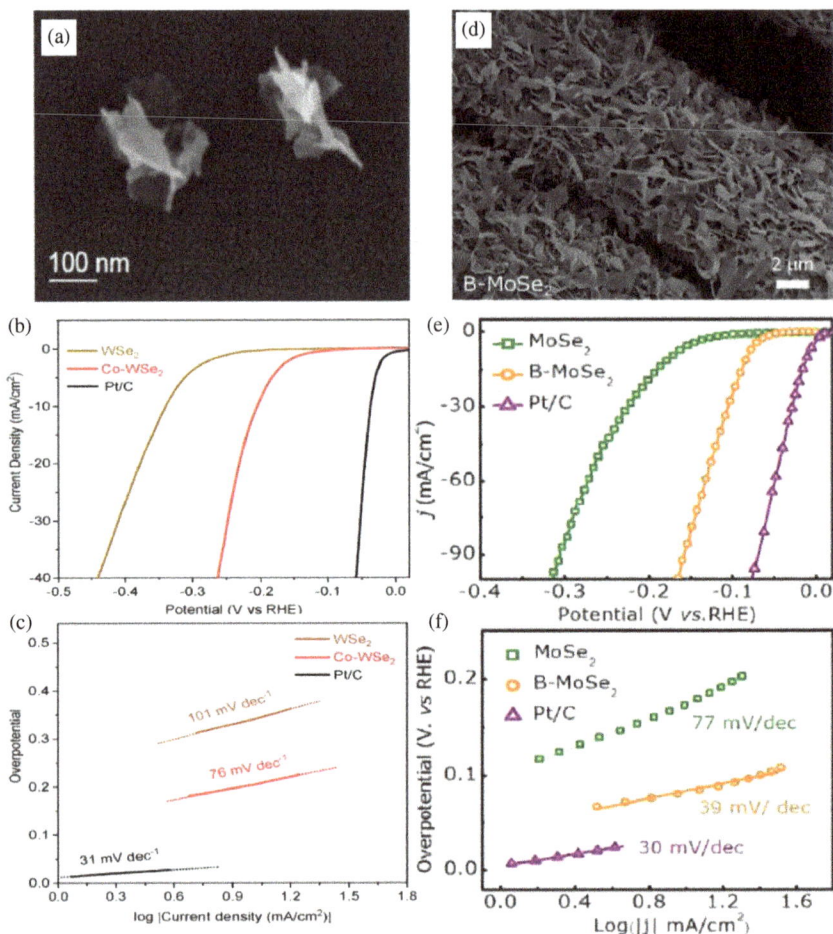

Fig 5.6   (a) TEM image of Co-WSe$_2$ nanosheets, (b) LSV curves, and (c) Tafel slope for Pt/C catalyst. [Reprinted with permission from Ref. [Zhang *et al.*, 2021] Copyright (2023) American Chemical Society]. (d) SEM image of B-doped Zhang MoSe$_2$ nanosheets, (e) LSV curves, and (f) Tafel slope for Pt/C catalyst (as a reference), MoSe$_2$ and B-doped MoSe$_2$ nanosheets in 0.5 M H$_2$SO$_4$. [Reprinted with permission from Ref. [Gao *et al.*, 2018]. Copyright (2023) Royal Society of Chemistry].

It implies that the minimum number of Co atoms gives the best hydrogen production performance. Fig 5.6(c) shows that the Tafel slope value of Co-WSe$_2$ is ~76 mV dec^{-1}, which is lower than the WSe$_2$ nanosheets (101 mV dec^{-1}). Tafel slope indicates the kinetics of electrocatalyst for hydrogen production due to the rapid charge-transfer ability, the lowering of H^+ free energy and reaction rate is observed after the Co doping. EIS analysis states that the charge transfer resistance (416 Ω) of Co-WSe$_2$ is minimal compared to the WSe$_2$ nanosheets (1553 Ω), indicating the better charge transfer ability at the reaction time. Gao *et al.* [2018] successfully demonstrated vertically aligned MoSe$_2$ flakes grown over carbon paper via the chemical vapor deposition (CVD) method by doping the B atom, activating the Se edge sites and inert basal plane. Authors suggested that computational study reveals the B-doping effect reduces the MoSe$_2$ formation energy and activates the electrocatalytic behavior of MoSe$_2$ for hydrogen production. Fig 5.6(d) shows an SEM image of B-doped MoSe$_2$ nanosheets, uniformly distributed over the carbon fiber. In Fig 5.6(e), the overpotential of B (5.5%)-doped MoSe$_2$ has a lower potential of ~84 mV compared with the pristine MoSe$_2$ (175 mV) at 10 mV cm^{-2}, which indicate the fast HER kinetics. The authors compared B-doped MoSe$_2$ with commercial Pt/C with 71 mV overpotential. The B-doped MoSe$_2$ shows a low Tafel slope value of 39 mV dec^{-1} compared to pristine MoSe$_2$ (77 mV dec^{-1}). The low Tafel value for B-doped MoSe$_2$ suggests the Volmer–Tafel process for hydrogen production when compared with Pt (30 mV dec^{-1}) in Fig 5.6(f). It is stable after 10,000 cycles, showing its robustness and long-term durability in 0.5 M H$_2$SO$_4$ electrolyte.

The chronoamperometry durability experiments were carried out at 100-mV overpotential and showed the current density remained stable for 20 hours. The B-doped MoSe$_2$ has a much smaller resistance (19 Ω), showing higher conductivity and faster electron movement. The B-doped electrocatalyst also showed high double-layer capacitance (C_{dl}) for 21 mF cm^{-2}, which is five times higher than pristine MoSe$_2$ nanosheets (4 mF cm^{-2}). All enhanced performance correlates to the doping effect.

Heterostructures of TMDs and carbon-based materials have shown promising results for HER. This layered structure of TMDs provides a high surface area and rich active sites for the HER. On the other hand, carbon materials in heterostructure offer high electrical conductivity and excellent stability. It is possible to achieve high synergistic effects that enhance the electrocatalytic activity and stability of heterostructure for HER by combining TMDs and carbon nanomaterials. Yang *et al.* [2018] demonstrated the synthesis of CNT-MoSe$_x$S$_{2-x}$ (vertical 2D molybdenum sulfoselenide nanosheet arrays over CNT) through a solid-state sulfidation reaction process to boost HER activity. In this study, the authors demonstrated that CNT-MoSe$_{1.06}$S$_{0.94}$ exhibits a low overpotential (η_{10}) of 174 mV (Tafel slope ~40 mV dec^{-1}) compared to CNT-MoSe$_2$ (η_{10} ~213 mV and Tafel slope ~46 mV dec^{-1}) and MoS$_2$-CNT (η_{10} ~224 mV and Tafel slope ~52 mV dec^{-1}). The high electrocatalytic performance of CNT-MoSe$_{1.06}$S$_{0.94}$ is due to the highly conductive CNTs acting as a backbone, which allows a uniform dispersion of molybdenum sulfoselenide nanosheets without aggregation. This results in an abundance of accessible active sites for subsequent electrocatalysis. The interconnected CNT network facilitates electron transport from active sites to current collectors. Liu *et al.* [2018] demonstrated porous graphene (P-rGO) using CO$_2$ as an activating agent and incorporated with chemically exfoliated 1T MoS$_2$ and formed a composite of MoS$_2$/P-rGO. The support of rGO with 1T MoS$_2$ enhances the electrochemical catalytic activity due to the porous nature, high electrical conductivity, and assessable surface area of rGO. The HER performance of rGO and p-rGO is weak. The onset potential of rGO and p-rGO decreases, which shows the enhancement in HER performance after the CO$_2$ activation. 1T MoS$_2$ shows good hydrogen production and an enhanced performance after incorporating the rGO (add 5 wt%), which decreases the overpotential (220 mV) and enhance the current density. The 1T MoS$_2$/P-rGO enhances the HER activity due to the high electrical conductivity and maximum number of active sites.

The Tafel slope of 89 and 100 mV dec^{-1} for 1T MoS$_2$ and 1T MoS$_2$/rGO shows a better performance. The small Tafel slope value (75 mV dec^{-1}) of 1T MoS$_2$/P-rGO as compared to the Pt electrode (30 mV dec^{-1}) shows the best electrocatalyst for hydrogen production. Zheng *et al.* [2020] produced defect-rich MoSe$_2$ with nitrogen-doped reduced graphene oxide (MoSe$_2$/NG) through a straightforward hydrothermal process. The SEM (Fig 5.7(a)) and TEM (Fig 5.7(b)) images of MoSe$_2$/NG-4 (Mo and Se molar ratio of 1: 4) clearly show

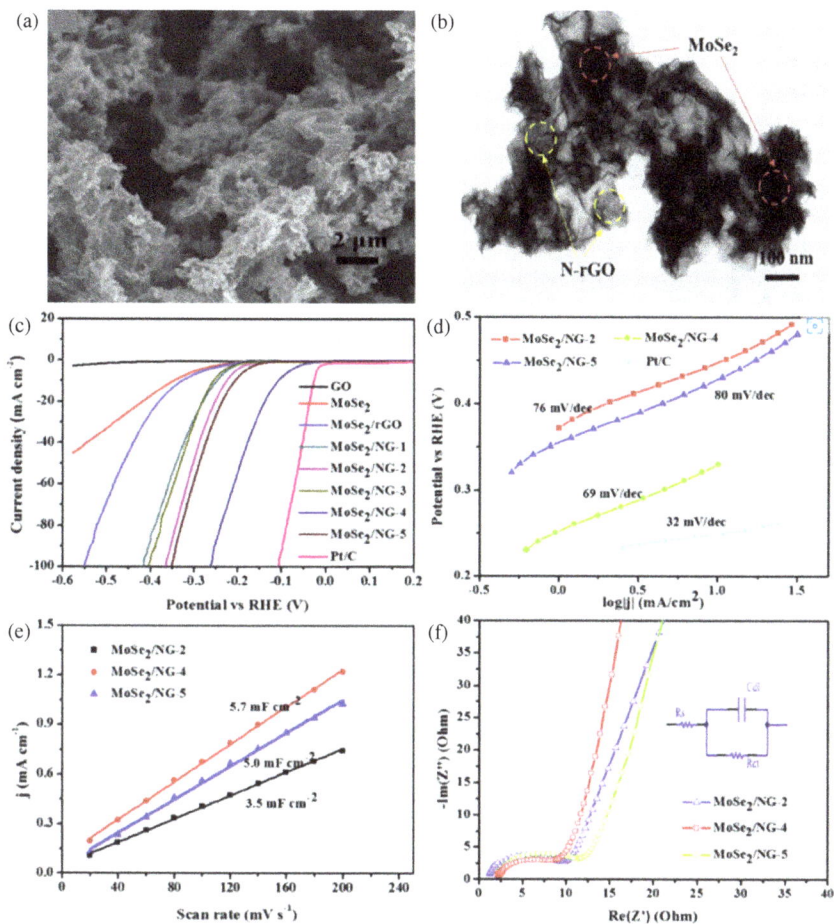

Fig 5.7 (a) SEM, (b) TEM images of MoSe$_2$/NG-4, (c) LSV curves, (d) corresponding Tafel plot at potential sweep rate of 5 mV s^{-1}, (e) current density vs. different potential sweep rates, and (f) EIS spectra of different samples. [Reprinted with permission from Ref. [Zheng *et al.*, 2020]. Copyright (2023) Elsevier].

that nanoflower morphology of MoSe$_2$ is homogeneously anchored on the nitrogen-doped graphene sheets in the prepared heterostructure. The catalyst MoSe$_2$/NG-4 exhibits a low overpotential of 120 mV at 10 mA cm^{-2} for the recorded LSV curve at a potential sweep rate of 5 mV s^{-1}, as shown in Fig 5.7(c). It also exhibits a low Tafel slope of

69 mV dec^{-1}, which is much lower than the values of other prepared samples, as shown in Fig 5.7(d). Fig 5.7(e) shows the plots for current density vs. potential sweep rates using cyclic voltammetry. It reveals that the catalyst MoSe$_2$/NG-4 shows high C$_{dl}$ (~5 mF cm^{-2}) indicating a higher electrochemically active surface area compared to other materials. Fig 5.7(f) shows the EIS spectra for prepared samples, and MoSe$_2$/NG-4 shows low charge-transfer resistance of 6.5 Ω. The synergistic effect of nitrogen doping, the good conductivity of graphene, and the abundance of active sites in heterostructure of MoSe$_2$/NG-4, are the prime reasons for its exceptional electrocatalytic property.

5.6 Conclusion

Green hydrogen gas was considered one of the most promising sustainable and clean energy sources for ecological systems. Electrolysis is the only technique of hydrogen generation that produces hydrogen that is 99.9% pure with zero emission of CO$_2$ gas compared with other production methods. This technique has attracted much interest due to its cost-effectiveness, environmental friendliness, high efficiency, and, more importantly, zero CO$_2$ emission. Nonmetal catalysts such as metal oxide, metal dichalcogenides, and carbon-based nanostructure are needed to reduce overpotential values and increase the efficiency and rate of reaction. The development of electrocatalysts can contribute to the solution of the problem of Pt depletion and the decrease in the energy efficiency and progress of efficient and low-cost, environmentally friendly catalysts. This chapter discusses the development of carbon- and TMD-based catalysts for the HER. Carbon-based materials such as CNTs and reduced graphene oxide have emerged as promising carbon-based electrode materials for HER. The catalytic activity of CNTs depends on their structure and surface termination. Cobalt embedded in nitrogen-rich CNTs (Co-NRCNTs) was developed as an effective electrocatalyst for HER, exhibiting activities similar to platinum. The 2D TMDs such as MoS$_2$, MoSe$_2$, WS$_2$, and WSe$_2$ are also used as electrocatalysts for hydrogen generation due to their high acidic stability, excellent electrochemical activity, and abundance in nature. Moreover, the crystal structure of catalysts plays a crucial role in determining their catalytic properties. The active edge sites of MoS_2 layers were found to contribute significantly to the catalytic activity for HER, despite their small surface area compared to the basal planes. Mechanically activated, distorted MoS_2 nanostructures and ultrathin WS_2 nanoflakes to maximize the exposure of active edge sites, resulting in improved HER performance. Combining TMDs and carbon-based materials in heterostructures has shown remarkable results for HER. TMDs such as $MoSe_2$ and WS_2 provide high surface area and numerous active sites, while carbon materials offer high electrical conductivity and stability. The synergistic effects achieved in TMD–carbon heterostructures enhance the electrocatalytic activity and stability of HER, making them attractive candidates for energy applications.

References

Ayusheev, A. B., Taran, O. P., Seryak, I. A., Podyacheva, O. Y., Descorme, C., Besson, M., Kibis, L. S., Boronin, A. I., Romanenko, A. I., Ismagilov, Z. R. & Parmon, V. (2014). Ruthenium nanoparticles supported on nitrogen-doped carbon nanofibers for the catalytic wet air oxidation of phenol. *Appl. Catal. B* 146, pp. 177–185.

Bae, G., Youn, D. H., Han, S. & Lee, J. S. (2013). The role of nitrogen in a carbon support on the increased activity and stability of a Pt catalyst in electrochemical hydrogen oxidation. *Carbon* 51, pp. 274–281.

Benck, J. D., Chen, Z., Kuritzky, L. Y., Forman, A. J. & Jaramillo, T. F. (1916). Amorphous molybdenum sulfide catalysts for electrochemical hydrogen production: Insights into the origin of their catalytic activity. *ACS Catal.* 2(9), pp. 1916–1923.

Benck, J. D., Hellstern, T. R., Kibsgaard, J., Chakthranont, P. & Jaramillo, T. F. (2014). Catalyzing the hydrogen evolution reaction (HER) with molybdenum sulfide nanomaterials. *ACS Catal.* 4(11), pp. 3957–3971.

Castro, E. B., De Giz, M. J., Gonzalez, E. R. & Vilche, J. R. (1997). An electrochemical impedance study on the kinetics and mechanism of the hydrogen evolution reaction on nickel molybdenite electrodes. *Electrochim. Acta* 42(6), pp. 951–959.

Chen, Z., Liu, X., Shen, T., Wu, C., Zu, L. & Zhang, L. (2021). Porous NiFe alloys synthesized via freeze casting as bifunctional electrocatalysts for oxygen and hydrogen evolution reaction. *Int. J. Hydrog. Energy* 46(76), pp. 37736–37745.

Cheng, L., Huang, W., Gong, Q., Liu, C., Liu, Z., Li, Y. & Dai, H. (2014). Ultrathin WS_2 nanoflakes as a high-performance electrocatalyst for the hydrogen evolution reaction. *Angew. Chem. Int. Ed.* 53(30), pp. 7860–7863.

Crabtree, G. W., Dresselhaus, M. S. & Buchanan, M. V. (2004). The Hydrogen Economy. *Phys. Today* 57(12), pp. 39–44.

Danilovic, N., Subbaraman, R., Strmcnik, D., Stamenkovic, V. R. & Markovic, N. M. (2013). Electrocatalysis of the HER in acid and alkaline media. *J. Serb. Chem. Soc.* 78(12), pp. 2007–2015.

Duan, J., Chen, S., Jaroniec, M. & Zhang Qiao, S. (2015). Heteroatom-doped graphene-based materials for energy-relevant electrocatalytic processes. *ACS Catal.* 5(9), pp. 5207–5234.

Gao, D., Xia, B., Zhu, C., Du, Y., Xi, P., Xue, D., Ding, J. & Wang, J. (2018). Activation of the MoSe2 basal plane and Se-edge by B doping for enhanced hydrogen evolution. *J. Mater. Chem. A* 6(2), pp. 510–515.

Holmes-Gentle, I. & Hellgardt, K. (2018). A versatile open-source analysis of the limiting efficiency of photo electrochemical water-splitting. *Sci. Rep.* 8(1), p. 12807.

Hu, C., Paul, R., Dai, Q. & Dai, L. (2021). Carbon-based metal-free electrocatalysts: From oxygen reduction to multifunctional electrocatalysis. *Chem. Soc. Rev.* 50(21), pp. 11785–11843.

Huang, H., Yan, M., Yang, C., He, H., Jiang, Q., Yang, L., Lu, Z., Sun, Z., Xu, X., Bando, Y., Yamauchi, Y., Huang, H. J., Yan, M. M., Yang, C. Z., He, H. Y., Jiang, Q. G., Yang, L., Lu, Z. Y., Sun, Z. Xu, X., Bando, Y. & Yamauchi, Y. (2019), Graphene nanoarchitectonics: Recent advances in graphene-based electrocatalysts for hydrogen evolution reaction. *J. Adv. Mater.* 31(48), p. 1903415.

Huld, T. & Gracia Amillo, A. M. (2015). Estimating PV module performance over large geographical regions: The role of irradiance, air temperature, wind speed and solar spectrum. *Energies* 8(6), pp. 5159–5181.

Jaramillo, T. F., Jørgensen, K. P., Bonde, J., Nielsen, J. H., Horch, S. & Chorkendorff, I. (2007). Identification of active edge sites for electrochemical H_2 evolution from MoS_2 nanocatalysts. *Science* 317(5834), pp. 100–102.

Kim, J., Kim, H. & Ahn, S. H. (2019). Electrodeposited rhodium phosphide with high activity for hydrogen evolution reaction in acidic medium. *ACS Sustain. Chem. Eng.* 8(6), pp. 5159–5181.

Li, F., Jiang, J., Wang, J., Zou, J., Sun, W., Wang, H., Xiang, K., Wu, P. & Hsu, J. P. (2023). Porous 3D carbon-based materials: An emerging platform for efficient hydrogen production. *Nano Res.* 16(1), pp. 127–145.

Liang, X. & Wu, C. M. L. (2020). Metal-free two-dimensional phosphorus carbide as an efficient electrocatalyst for hydrogen evolution reaction comparable to platinum. *Nano Energy* 71, pp. 104603.

Liu, Y., Liu, J., Li, Z., Fan, X., Li, Y., Zhang, F., Zhang, G., Peng, W. & Wang, S. (2018). Exfoliated MoS_2 with porous graphene nanosheets for enhanced electrochemical hydrogen evolution. *Int. J. Hydrog. Energy* 43(30), pp. 13946–13952.

Lu, Q., Yu, Y., Ma, Q., Chen, B. & Zhang, H. (2016). 2D Transition-metal-dichalcogenide-nanosheet-based composites for photocatalytic and electrocatalytic hydrogen evolution reactions. *Adv. Mater.* 28(10), pp. 1917–1933.

Ma, L., Shen, X., Zhou, H., Zhu, G., Ji, Z. & Chen, K. (2015). CoP nanoparticles deposited on reduced graphene oxide sheets as an active electrocatalyst for the hydrogen evolution reaction. *J. Mater. Chem. A* 3(10), pp. 5337–5343.

Mishra, S., Jaiswal, S. S. & Mishra, A. K. (2022). Multiwalled carbon nanotubes grown over green iron nanocatalyst as electrode for hydrogen-producing electrochemical cell. *J Mater. Sci. Mater Electron.* 33(11), pp. 8702–8710.

Mishra, S. & Mishra, A. K. (2020). Hydrothermally synthesized MoS_2 nanoclusters for hydrogen evolution reaction. *Electroanalysis* 32(11), pp. 2564–250.

Moca, R. (2019). Novel inorganic material for hydrogen evolution reaction in electrochemical water (Doctoral dissertation, University of Glasgow).

Nie, R., Miao, M., Du, W., Shi, J., Liu, Y. & Hou, Z. (2016). Selective hydrogenation of CC bond over N-doped reduced graphene oxides supported Pd catalyst. *Appl. Catal. B* 180, pp. 607–613.

Parsons, R. (1958). The rate of electrolytic hydrogen evolution and the heat of adsorption of hydrogen. *J. Chem. Soc. Faraday Trans.* 54, pp. 1053–1063.

Rajesh Kumar, C. J. & Majid, M. A. (2020). Renewable energy for sustainable development in India: Current status, future prospects, challenges, employment, and investment opportunities. *Energy Sustain. Soc.* 10(1), pp. 1–36.

Shin, S., Jin, Z., Kwon, D. H., Bose, R. & Min, Y. S. (2015). High turnover frequency of hydrogen evolution reaction on amorphous MoS2 thin film directly grown by atomic layer deposition. *Langmuir* 31(3), pp. 1196–1202.

Singh, A., Majee, B. P., Gupta, J. D. & Mishra, A. K. (2023). Layer dependence of thermally induced quantum confinement and higher order phonon scattering for thermal transport in CVD-grown triangular MoS_2. *J. Phys. Chem. C* 127(7), pp. 3787–3799.

Skúlason, E., Karlberg, G. S., Rossmeisl, J., Bligaard, T., Greeley, J., Jónsson, H. & Nørskov, J. K. (2007). Density functional theory calculations for the hydrogen evolution reaction in an electrochemical double layer on the Pt(111) electrode. *Phys. Chem. Chem. Phys.* 9(25), pp. 3241–3250.

Tributsch, H. & Bennett, J. C. (1977). Electrochemistry and photochemistry of MoS_2 layer crystals. I. *J. Electroanal. Chem.* 81(1), pp. 97–111.

Wang, D., Wang, Z., Wang, C., Zhou, P., Wu, Z. & Liu, Z. (2013). Distorted MoS_2 nanostructures: An efficient catalyst for the electrochemical hydrogen evolution reaction. *Electrochem. Commun.* 34, pp. 219–222.

Wu, M., Liao, J., Yu, L., Lv, R., Li, P., Sun, W., Tan, R., Duan, X., Zhang, L., Li, F., Kim, J., Ho Shin, K., Seok Park, H., Zhang, W., Guo, Z., Wang, H., Tang, Y., Gorgolis, G. & Galiotis, C. (2020). 2020 Roadmap on carbon materials for energy storage and conversion. *Chem. Asian J.* 15(7), pp. 995–1013.

Yang, J., Liu, Y., Shi, C., Zhu, J., Yang, X., Liu, S., Li, L., Xu, Z., Zhang, C. & Liu, T. (2018). Carbon nanotube with vertical 2D molybdenum sulphoselenide nanosheet arrays for boosting electrocatalytic hydrogen evolution. *ACS Appl. Energy Mater.* 1(12), pp. 7035–7045.

Yuan, S., Pang, S. Y. & Hao, J. (2020). 2D transition metal dichalcogenides, carbides, nitrides, and their applications in supercapacitors and electrocatalytic hydrogen evolution reaction. *Appl. Phys. Rev.* 7(2), p. 021304.

Zhang, W., Liu, X., Liu, T., Chen, T., Shen, X., Ding, T., Cao, L., Wang, L., Luo, Q. & Yao, T. (2021). In situ investigation on doping effect in co-doped tungsten diselenide nanosheets for hydrogen evolution reaction. *J. Phys. Chem. C* 125(11), pp. 6229–6236.

Zhang, G., Zhao, D., Yan, J., Jia, D., Zheng, H., Mi, J. & Li, Z. (2019). The promotion and stabilization effects of surface nitrogen containing groups of CNT on cu-based nanoparticles in the oxidative carbonylation reaction. *Appl. Catal. A: Gen.* 579, pp. 18–29.

Zheng, D., Cheng, P., Yao, Q., Fang, Y., Mei, Y., Zhu, L. & Zhang, L. (2020). Excess Se-doped $MoSe_2$ and nitrogen-doped reduced graphene oxide composite as electrocatalyst for hydrogen evolution and oxygen reduction reaction. *J. Alloys Compd.* 848, p. 156588.

Zou, X., Huang, X., Goswami, A., Silva, R., Sathe, B. R., Mikmeková, E. & Asefa, T. (2014). Cobalt-embedded nitrogen-rich carbon nanotubes efficiently catalyze hydrogen evolution reaction at all pH values. *Angew. Chem. Int. Ed.* 53(17), pp. 4372–4376.

Zou, X. & Zhang, Y. (2015). Noble metal-free hydrogen evolution catalysts for water splitting. *Chem. Soc. Rev.* 44(15), pp. 5148–5180.

Chapter 6

Carbon-Based and TMDs-Based Nanostructures for Solar-Driven Steam Generation

Higgins M. Wilson and Neetu Jha

Department of Physics, Institute of Chemical Technology, Mumbai
Nathalal Parekh Marg, Mumbai-400019 India
Email: nr.jha@ictmumbai.edu.in

Abstract

This chapter discusses the usage of carbon- and transition metal dichalcogenide (TMD)-based materials for Solar-driven steam generation. Carbon-based materials absorb highly in the whole range of solar spectrum and TMD materials possess high surface plasmonic resonance. This prompts the scientific community to employ them as photothermal materials for solar evaporation. This principle has been utilized for speeding up the process of distillation and hence for the modification of commercial solar still. Photothermal material membrane–modified solar still can purify water systems including seawater, industrial wastewater, and sewage water. This chapter deals with the discussion of various types of photothermal materials, their photothermal efficiencies, and different modifications of the floating systems to reduce the thermal losses in the system.

6.1 Introduction

In the past decade, solar energy has been touted as a sustainable solution for the impending global energy shortage. The rise in the world population, coupled with rapid industrialization and modernization, has necessitated the development of renewable sources of energy capable of replacing the rapidly decreasing exhaustible energy resources. Among the various available inexhaustible sources of energy, solar energy has gained the most traction, owing to its considerable untapped potential [Gueymard, 2004].

Apart from energy shortage, the availability of freshwater is also a growing concern, owing to increasing global demand with diminishing natural freshwater resources. In the current global scenario, numerous water purification techniques, such as desalination, reverse osmosis, and distillation, have been employed to meet the freshwater demand [Anderson *et al.*, 2010]. However, detrimental factors such as cumbersome purification setups, high cost, and external energy requirements render these techniques unsuitable as a sustainable long-term solution. Distillation is a common separation technique to purify water and has been utilized since ancient times. It involves the heating of water to its boiling point to induce phase change, and its subsequent condensation to obtain clean water. Despite its simplicity, the large-scale application of distillation in water purification is unsuitable, owing to the requirement of a constant supply of heat. The heat can be supplied by burning fuel, utilizing electricity, or by focusing sunlight on water to produce steam [Deng *et al.*, 2017]. In industries, steam is crucial for various industrial operations and is generated by heating bulk water by burning coal or concentrating solar illumination. However, the rapid decline of fossil fuels and the perilous environmental ramifications of burning fossil fuels make fossil-fuel-based steam generation unsustainable for future operations. Hence, solar-driven steam generation has attracted interest among researchers around the globe owing to its potential as a long-term sustainable solution for steam generation.

Despite the obvious advantages of solar-driven steam generation, its applications in current large-scale systems have been scarce. This can be attributed to the requirement of large optical solar concentrators and complex support structures to focus solar illumination. In the currently used solar stills without focused illumination, the solar flux (<1000 W m^{-2}) is insufficient to overcome the latent heat of vaporization of water for constant steam generation. Therefore, parabolic troughs, heliostats, and other optical concentrators are crucial for achieving elevated temperatures at the point of focus [Gao *et al.*, 2019]. However, the high price of optical concentrators accounts for a major percentage of the capital cost. Furthermore, the electricity required for driving solar-tracking systems and the poor photothermal performance of the solar concentrators hinder the real-life applications of these solar thermal systems.

Plasmonic nanoparticles (NPs) have recently emerged as a promising solution for absorbing solar irradiation in water-based nanofluids (NFs), utilizing surface plasmon resonance to generate hot electrons. However, NF-based solar steam generation has faced challenges in achieving significant photothermal performance improvements. It often requires 10–1000 times higher optical solar concentration levels for effective steam generation. In addition, issues such as inadequate NF dispersion, inefficient absorption due to scattering, and NP presence on the surface have hindered their practical applicability. To overcome these limitations, a novel approach has been developed. It involves the design of a floating absorber with broad absorbance, excellent hydrophilicity, and low thermal conductivity. This innovative technique allows the absorber to float on the water's surface and efficiently promotes interfacial evaporation at the air–water interface. By doing so, it effectively mitigates the optical losses experienced in NF-based solar steam generation systems, leading to improved performance and practical feasibility.

This chapter provides a comprehensive overview of recent advancements in structural design for solar-induced evaporation systems, emphasizing their ability to achieve remarkably high photothermal efficiencies. The applications of these systems are also explored in detail. Initially, the phenomenon of solar-driven steam generation and its influential parameters are briefly reviewed. Subsequently, the focus shifts to the progress made in solar steam generation designs, particularly in interfacial evaporation systems that attain high photo-thermal efficiencies even under low solar flux conditions. The utilization of solar-induced photothermal effects in various applications, such as seawater desalination, water purification, dye degradation, electricity generation, and more, is thoroughly discussed. Ultimately, a perspective is provided on the diverse applications of solar-induced evaporation designs, aiming to guide researchers in material designs for high photothermal performance in steam generation and provide direction for their implementation in different research fields. This review aims to offer valuable insights into recent trends and serve as a reference for designing interfacial solar steam generation systems, fostering advancements across diverse disciplines.

6.2 Light-to-Heat Conversion Mechanisms

Solar energy can easily be converted into other forms of energy, such as electrical and thermal, using existing technology. However, owing to their low efficiencies, solar-driven energy systems are incapable of replacing fossil fuels as a primary source of energy. Considerable research is being done in photovoltaics to increase the efficiency of solar cells and obtain efficiencies closer to their theoretical limit. Other recent applications of solar energy, such as solar-driven water splitting for hydrogen gas production are still at a laboratory scale. Moreover, the maximum possible efficiency of energy conversion is still meager compared to solar-thermal-conversion-based systems, which boast the highest achievable conversion efficiency [Gao *et al.*, 2019]. Solar thermal or photothermal energy conversion occurs due to the photoexcitation of electrons, leading to heat generation. The photothermal effect can be observed in inorganic materials, such as plasmonic metals and semiconductors, and organic materials, such as carbonaceous materials and polymers. Plasmonic metallic NPs consist of a significant number of free electrons in the conduction band, according to Drude–Lorentz model [Deng *et al.*, 2017]. These free electrons undergo transitions to higher conduction band states upon illumination under certain wavelengths. These wavelengths are greater than the size of the NPs and play a crucial role in inducing localized surface plasmon resonance (LSPR). Moreover, the maximum absorption of light occurs at wavelengths that match the resonance frequency of the material.

In transition metal dichalcogenides (TMDs) and semiconductors, heat is generated via a non-radiative relaxation mechanism. A forbidden energy bandgap separates the conduction non-radiative and valence bands. When illuminated with an incident wavelength comparable to this bandgap energy, the valence electrons move to the conduction band leaving holes in their wake, thus forming electron–hole pairs [Almond *et al.*, 2020]. When the electron–hole pairs relax to lower energy levels, energy is released. The released energy is either in the form of photons after electron–hole recombination in the radiative decay process or in the form of phonons arising from the non-radiative process through thermalization. The generated phonons induce lattice heating, the heat from which is dispersed to the surrounding media.

Among organic materials, carbon-based materials and polymers are known to induce high photothermal effects [Gao *et al.*, 2019]. In carbon-based materials, photothermal heat is generated via lattice vibrations. Amorphous carbons with sp^2 and sp^3 bonding exhibit excellent broadband light absorption, along with graphitic structured carbons. When a carbon-based material is illuminated with energy equal to the electronic transition of the molecule, the transfer of electrons is induced from the highest occupied molecular state (HOMO) to the lowest unoccupied molecular state (LUMO) state. This is attributed to the presence of conjugated π bonds which undergo π to π* transitions. These transitions occur at low energies and hence can be induced by the incident solar illumination. The presence of a large number of conjugated π bonds aids in the broadband absorption of solar illuminations in graphene-like structures. The photoexcited electrons undergo photon-phonon interactions, which transfer energy to the vibrational modes of the molecules.

6.3 Parameters Influencing Solar-Driven Steam Generation

6.3.1 *Solar Absorptance*

Solar absorptance plays a crucial role in the photothermal capability of materials. Solar absorptance is defined as the capability of absorption of incident wavelengths by the material across the standard solar spectrum (AM 1.5G). Hence, the material should possess a high solar absorptance value (250–2500 nm) for efficient solar-driven steam generation across the solar spectrum. Furthermore, the reflectance and transmittance of the material also influence its performance. For excellent light absorption and heat conversion, the values of reflectance and transmittance should be minimum across the solar spectrum.

6.3.2 *Solar Thermal Conversion Efficiency*

In an ideal situation, a photothermal material should absorb most of the incoming solar illumination and efficiently convert it to heat energy. Moreover, the generated heat should be enough to overcome the latent heat of vaporization of water and produce steam. Therefore, the performance of solar thermal evaporation systems can be quantified using solar thermal conversion efficiencies. Solar thermal efficiency (η) can be expressed with the following equation:

$$\eta = \frac{m\,h_t}{I} \tag{6.1}$$

where m is the effective evaporation rate in kg m^{-2} h^{-1}, h_t is the total heat enthalpy for vaporization, and I is the incident flux of solar illumination [Gao *et al.*, 2019].

The effective evaporation rate is calculated after subtracting the dark evaporation rate from the water evaporation rate under incident illumination. The dark evaporation rate is the evaporation rate of water in solar thermal systems under dark or no-light conditions. For accurate measurements, the slope of the graph of mass change versus time is utilized in calculations. The total heat enthalpy ($h_t = h_{vap} + h_s$) is the total energy required for the occurrence of phase change and the evaporation of water at room temperature. This includes the latent heat of vaporization (h_{vap} = 2260 kJ kg^{-1} at 100°C) and sensible heat (h_s), which is the energy required to heat water from room temperature (t_1) to vaporization temperature (t_2). It can be defined as $h_s = 4.2 (t_2 - t_1)$.

6.4 Solar-Driven Steam Generation Techniques

6.4.1 *Conventional Solar Steam Generation*

Conventional solar steam generation (CSSG) is primarily utilized in industries. CSSG exhibits low photothermal conversion efficiencies under incident solar flux. This is attributed to the poor solar absorptance of water, which has a high absorbance in the near-IR region and reflects most of the solar radiation. The radiation comprises wavelengths of 250–400 nm (UV), 400–700 nm (VIS), and 700–2500 (IR). This high wastage of incident solar flux results in poor photothermal conversion and necessitates the use of solar concentrators like parabolic troughs or Fresnel reflectors for concentrating sunlight to generate sufficient heat for water evaporation.

Absorbers with high photothermal conversion efficiencies are being utilized in industrial applications where the incident flux is generally as high as 50 suns. However, they are not optimal, owing to the separation between the absorber area and the fluid. This separation induces temperature differences as high as 500°C [Abdelrahman *et al.*, 1979]. High temperatures lead to extremely high radiative losses because radiation losses are proportional to the fourth power of surface temperature. The heat-transfer fluids utilized in CSSG (e.g., ethylene glycol) usually exhibit poor solar absorption. Therefore, additives with high solar absorption (e.g., SiC and SiO_2) are added to the working fluids to improve their solar absorption. However, these additives are micron-sized particles that tend to agglomerate and settle in the solution. This, along with practical problems such as clogging, limits the application of additives in CSSG systems.

6.4.2 *Nanofluid-based Solar Steam Generation (NSSG)*

The limitations of macro-sized additives in CSSG systems can be alleviated with the use of nano-sized particles. Submicrometer particles in a solution are governed by two phenomena, that is, Brownian motion and gravitation. However, as the particle size decreases and nears the nanometer scale, Brownian motion becomes dominant, reducing the agglomeration of particles. Hence, scientists have focused on replacing the macro-sized additives with functionalized nanoparticle suspensions called nanofluids (NFs). These NFs contain NPs with sizes less than 100 nm and exhibit distinct advantages over traditional macro-sized fluids in thermal applications. The higher specific surface area of the NPs facilitates higher heat transfer between particles and fluids. Furthermore, the NFs are more stable, owing to the dominant Brownian motion in the nanometer scale. Stable NFs are also a requirement in industries to endure harsh environments like high temperatures. Finally, the size tunability of NPs aids in improving the thermal conductivity and surface wettability of NFs [Gao *et al.*, 2019].

The physical mechanism behind NSSG, commonly known as volumetric heating, has been debated in recent years (Fig 6.1). In the first proposed mechanism, the steam generation is attributed to the formation of nanobubbles around the nanoparticles in a nonequilibria manner at the air–liquid interface [Almond *et al.*, 2020]. In this mechanism, the temperature of the surrounding bulk fluid does not increase with a rise in the NP temperature. Moreover, steam is generated because of the formation of localized nanobubbles. For example, nanobubble formation was observed in Au NPs when illuminated with a laser. The nanobubbles formed at an intensity threshold of 3×10^8 suns. The second mechanism focuses on the NPs attaining equilibrium with the surrounding media. Furthermore, it attributes the formation of steam to the rise in bulk fluid temperature.

6.4.3 *Carbon Materials for NSSG*

Recently, carbon-based materials have been extensively studied for volumetric solar steam generation. Their popularity can be ascribed

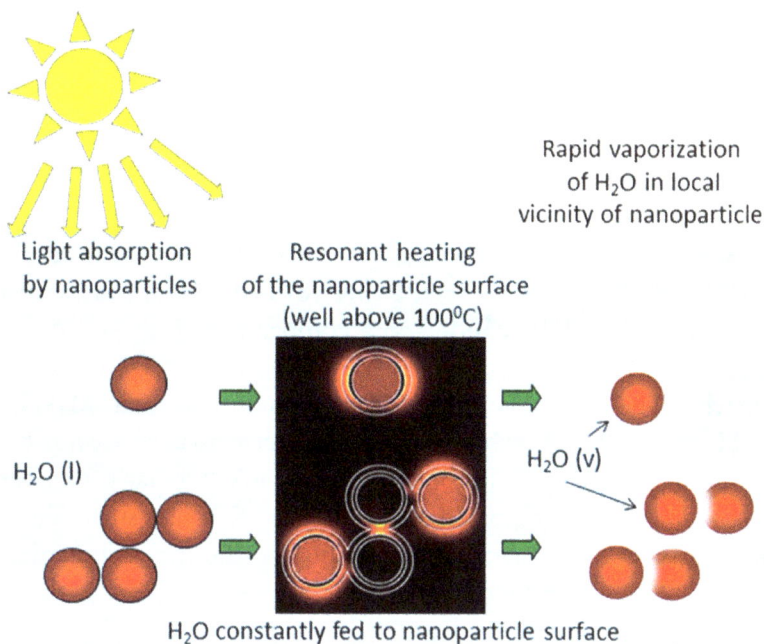

Fig 6.1. Nanobubble formation under solar illumination. Reprinted with permission from Neumann *et al.* [2013]. Copyright (2023) American Chemical Society.

to the following reasons: (1) They have a significantly lower cost compared to metallic NPs and (2) exhibit broadband absorbance in the solar spectrum. For example, Ni *et al.* [2015] compared the steam generation performances of graphitized carbon black (GCB), carbon black (CB), and graphene for volumetric steam. They achieved efficiencies of 69%, 68%, and 67% under illumination of 10 suns with CB, graphene, and GCB, respectively. However, during the testing for transient ($0 < t < 300$ s) evaporation, GCB exhibited the best performance. This was attributed to the low zeta potential of GCB (-40 mV). Lower zeta potential leads to lesser agglomeration and more stability in NFs. The study also elucidated the mechanism of vapor generation and utilized a lumped capacitance model to prove the rise in bulk temperature as the reason for vapor generation under low solar intensities ($0 < c < 10$). Furthermore, Ulset *et al.* [2018]

demonstrated CB NP-based NFs. They achieved 73% efficiency with 3 wt% solution under a parabolic concentrator with 0.76 sun in the Nordic region. Carbon nanotubes (CNTs), a carbonaceous material with high and broadband solar absorption, were reported by Wang *et al.* for volumetric steam generation [Wang *et al.*, 2016]. Single-walled CNTs (SWCNTs) of different concentrations were tested for steam generation and the highest efficiency (45%) was achieved under 10 sun illuminations with 19.04 vol% of CNTs. Furthermore, Ghafurian *et al.* [2019] conducted a comparative study of SWCNTs performance with functionalized MWCNTs (MWCNT-COOH) and hydroxyl MWCNTs (MWCNT-OH). They found that MWCNT-COOH of 0.002 wt% exhibited the best performance, owing to its good stability with a zeta potential of –35.5 mV. Similarly, graphene-based NFs have been reported in recent years for solar steam generation, owing to their lower cost and higher thermal conductivities compared to noble metals.

6.4.4 *Solar-Driven Interfacial Steam Generation (SISG)*

Despite the formulation of NFs for solar steam generation, their viability for application in practical systems remains uncertain. The undesirable bulk heating of the working fluid, poor photothermal efficiencies in low solar flux, and high thermal losses due to radiation, convection, and conduction are factors that contribute to this uncertainty. In the last few years, researchers have proposed a solar-driven interfacial evaporation technique for steam generation (SISG) to reduce the high thermal losses incurred in the volumetric heating of water. The SISG technique focuses on promoting interfacial evaporation at the air–water interface while simultaneously minimizing the bulk heating of water [Ghasemi *et al.*, 2014]. This is achieved with a floating absorber with excellent solar absorbance, high porosity, and low thermal conductivity. Despite the utilization of similar broadband absorbing photothermal materials in SISG, its steam generation mechanism differs significantly from volumetric heating. The 2D/3D designs enable the localization of the heat generated at the air–water interface. This localized heat aids in the rapid evaporation of the underlying water molecules in contact with the hot zone generated at the surface of the floating absorber. With proper thermal management, the hot zone can be maintained for long durations of time without heat loss to the underlying bulk water. This results in excellent water evaporation rates even at low solar flux [Gao *et al.*, 2019].

6.4.5 *Carbon-Based Photothermal Materials for SISG*

The first step in the development of solar steam generators using SISG technique is the selection of photothermal materials. Materials exhibiting broadband absorbance in the range of 250–2500 nm are ideal, owing to their capability to generate thermal energy from incident solar flux. The last few years have witnessed the exploration of various photothermal materials for high-performance solar steam generation with excellent photothermal efficiencies. These materials range from carbonaceous materials, such as graphene oxide (GO), graphene, soot, and activated carbon to inorganic materials, such as Au NPs, CuS films, MoS_2, and MXene.

Most carbonaceous materials are naturally black in color, which generally implies excellent solar absorptance. Therefore, numerous carbon-based materials, such as graphite, CB, graphene, CNTs, soot, and reduced GO (rGO) have been utilized as photothermal materials for interfacial evaporation.

Graphite-based: Ghasemi *et al.* [2014] demonstrated interfacial evaporation by utilizing exfoliated graphite, which absorbed 97% of the solar power incident on the top exfoliated layer. Moreover, it exhibited a meager reflectivity of 3% when analyzed using a spectrophotometer equipped with an integrating sphere.

GO-based: GO exhibits a broad optical absorption over the visible and near-infrared (NIR) regions of the electromagnetic spectrum and excellent photothermal transduction. A scalable GO film demonstrated an absorption efficiency of >94% [Li *et al.*, 2016].

rGO-based: Graphene has attracted attention in recent years for its broadband solar absorbance, low molar specific heat, high Debye temperature, and tunable thermal and electrical conductivities. A single layer of graphene exhibits an absorbance of only 3%, which can be further improved significantly by increasing the graphene monolayers. This establishes graphene as an excellent potential candidate for broadband solar absorption. Jiang *et al.* [2016] utilized rGO for solar absorption in rGO-bacteria nanocellulose (RGO/BNC: BNC) aerogel which exhibited extremely small optical transmittance (~1.5%) and reflectance (~2.5%) in the visible and near-IR region.

Carbon black–based: CB has been studied for SISG, owing to its broadband absorbance and low cost. CB was coated on cotton gauze and cotton fabric for efficient solar absorption [Liu *et al.*, 2015; Li *et al.*, 2015]. Jin *et al.* [2018] spin-coated CB on nylon fibers and obtained an absorption of 94%, which was close to the absorption of pure CB (95%). Hou *et al.* [2019] reported a functionalized CB (fCB) membrane with 96% absorption and a reflectivity of less than 3% in the 250–2500 nm region. Moreover, a 3% transparency was obtained in the wavelength range of 200–1500 nm.

Carbon nanotube (CNT)–based: CNTs are completely black with a high absorbance in the solar spectrum. Several interfacial solar steam generators (ISSGs) that utilize CNTs as a photothermal material have been reported in the literature, for instance, MWCNTs coated on top of a macroporous silica material by vacuum filtration [Wang *et al.*, 2016]. The light diffuse reflection and transmission of the CNT top layers (1.2 to 2.4 µm) within the UV and visible ranges exhibited an average diffuse reflection of less than 2% and almost zero transmission, indicating nearly complete light harvesting (Fig 6.2d–e).

Carbonized materials: Carbonization of carbon precursors at high temperatures can produce excellent broadband solar absorbers. Among natural precursors, high-temperature carbonization of wood has gained tremendous interest in recent years like alcohol flame-treated wood, carbonized basswood, carbonized pinewood, surface-carbonized longitudinal wood, and carbonized sintered saw dust for

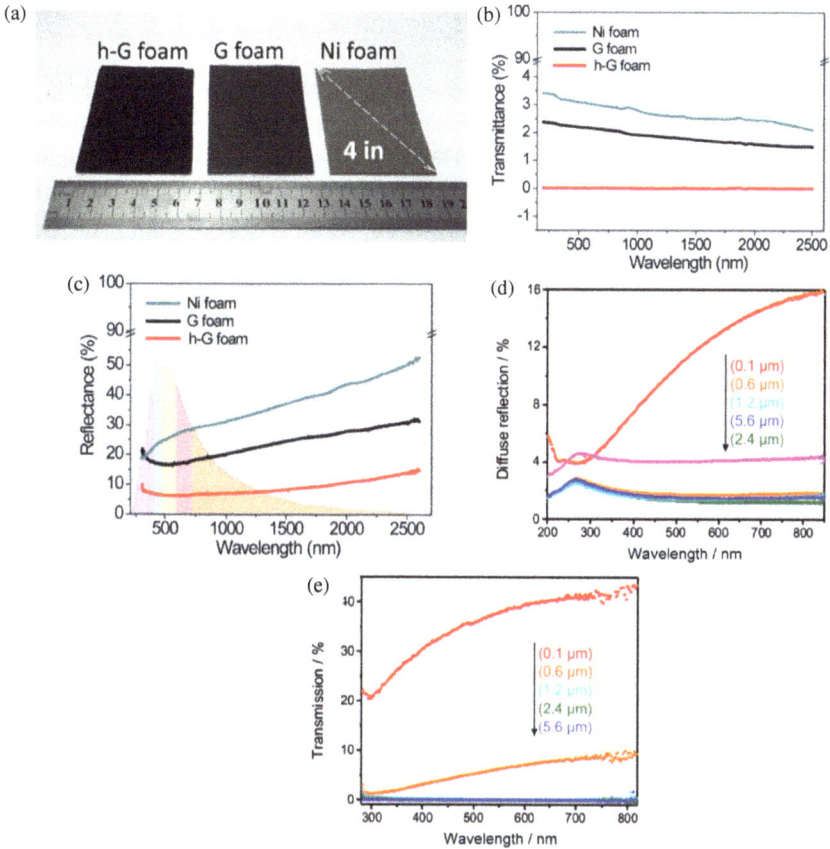

Fig 6.2 (a) Optical image, (b) transmittance, and (c) reflectance spectra of h-G foam, G foam, and Ni foam. Reprinted with permission from Ren *et al.* [2017]. Copyright (2020) John Wiley and Sons. (d) Diffuse reflection and (e) transmission spectra of CNT top layers from CNT-silica bilayered materials. Reprinted with permission from Wang *et al.* [2016] Copyright (2023) American Chemical Society.

SISG due to their tremendous light absorption (95–99%) in solar spectrum [Xue *et al.*, 2017].

Soot-based: Soot is another low-cost, scalable carbonaceous material with broadband absorbance. Authors utilized candle soot on cotton fabric and nylon for broadband optical absorption [Zhang *et al.*, 2018; Wilson *et al.*, 2019].

6.4.6 *TMD-based Materials for ISSG*

TMDs have gained attention due to their exceptional light absorption properties in the visible to near-infrared (vis–NIR) range. Monolayer molybdenum disulfide (MoS_2) has been found to exhibit better light absorption (5–10% of incident sunlight) compared to graphene (2.3%) and outperformed widely used solar absorbers such as GaAs and Si [Tsai *et al.*, 2014]. In addition, the narrow bandgap of MoS_2 (1.2–1.9 eV) allows for efficient heat generation through nonradiative decay of electron–hole pairs. Jun *et al.* explored the use of chemically exfoliated MoS_2 (ce-MoS_2) for solar distillation [Ghim *et al.*, 2018]. They achieved enhanced light absorption by inducing a phase transition from the 2H to 1T phase of MoS_2 through Li-intercalation, which narrowed the bandgap. By using bacterial nanocellulose (BNC) as a hydrophilic and thermally insulating polymer matrix, they fabricated a ce-MoS_2/BNC solar evaporator with significantly higher solar evaporation efficiency (6.15 kg $m^{-2}h^{-1}$) compared to bulk MoS_2/BNC (4.73 kg m^{-2} h^{-1}) under the same irradiation conditions. A three-dimensional MoS_2 aerogel demonstrated a light absorption efficiency of over 95% across the solar spectrum [Wang *et al.*, 2020]. The evaporation rate of the aerogel exceeded 90% under sunlight conditions with an intensity of 1.5–3.0 kW m^{-2}. Other MoS_2-based solar evaporation systems, such as MoS_2 nanoflower–based solar evaporators, double-layer MoS_2@sponge, and PEGylated MoS_2-cotton cloth, have also been developed for water evaporation and wastewater treatment [Liu *et al.*, 2022].

Compared to binary (transition metal chalcogenides) TMCs such as MoS_2, ternary TMCs such as $CuFeS_2$ and $CuFeSe_2$ offer greater flexibility in functional regulation and applications. Hu's team loaded narrow-bandgap $CuFeSe_2$ nanoparticles onto a wooden substrate for desalination, achieving high solar thermal efficiency (86.2%) under 5 kW m^{-2} and a solar light absorption of 99% across the full spectral range [Liu *et al.*, 2018]. Indirect-bandgap TMCs such as SnSe and $SnSe_2$ have high photothermal conversion capacity due to efficient phonon generation and low thermal conductivity. However, their limited light absorption across the full solar spectrum has hindered their application in solar evaporation. Jang *et al.* addressed this issue by depositing a large-area SnSe-SnSe$_2$ layer on a glassy carbon form (CF) using pulsed laser deposition (PLD) to construct a bilayer solar steam generator (BSSG) [Tahir *et al.*, 2020]. The porous CF trapped the incident light, leading to maximized light absorption and achieving an evaporation efficiency of 84.1% under one sun. In summary, TMCs, including MoS$_2$, CuFeS$_2$, CuFeSe$_2$, SnSe, and SnSe$_2$, exhibit unique properties that make them suitable for solar absorption and evaporation applications. Ongoing research and development in this field, as highlighted by the cited studies, hold promise for advancing solar energy utilization and addressing various environmental and technological challenges.

6.4.7 *Structural Configurations of SISG*

The next step in the design of SISG is the incorporation of photothermal materials in floating substrates. In this regard, 2D and 3D structures have been reported extensively for SISG. In 2D configurations, the photothermal materials are either designed into self-floating films or incorporated into floating substrates such as paper, wood, cloth, and so forth. The two primary factors which should be considered while selecting suitable substrates are thermal conductivity and hydrophilicity which impact the thermal management and water transportation of the evaporation system. In the initial configurations of 2D-structured evaporation systems, the conduction loss of heat from the localized air–water interface to the underlying bulk water decreased the overall photothermal performance of the system. Therefore, an extra insulation layer was provided between the 2D structure and underlying bulk water to negate the downward conduction loss of heat in future evaporation systems. For example, Li *et al.* [2016] introduced polystyrene foam between GO film (photothermal layer) and bulk water and obtained 80% photothermal efficiency under 1 sun illumination. The conduction loss of GO film without a thermal insulator was reported to be as high as 50% with 1-mm film thickness. The percent loss, however, decreased with increasing thickness of GO film which reiterates the need for separation of heat localization zone at the air–water interface from underlying bulk water. Meanwhile, the surface temperature of the photothermal absorber is low and close to room temperature in SISG due to effective evaporation. This low-temperature differential helps in decreasing the radiation and convection losses of heat to the surrounding media. Moreover, the convection loss of heat due to the test container containing the bulk fluid surrounding shall also contribute to the decrease in the evaporation performance of the system as demonstrated by Ghasemi *et al.* [2014]. Hence, double concentric acrylic tubes filled with hydrophobic aerogels were utilized to minimize the 6.5% (3% top walls, 3.5% side walls) projected convection loss to the surroundings. Despite the excellent thermal management and evaporation performance of 2D systems, the lack of adequate thermal energy storage contributes to inevitable heat losses upon continuous operation. 3D systems, however, mitigate this issue with intrinsic water supply channels and added thermal insulation. In addition, 3D structures also have better heat absorption capabilities, vapor escape channels, and water transportation networks which enable efficient long-term steam generation. To date, various 3D structures such as foams, sponges, and gels have been reported for SISG with photothermal efficiencies going over the theoretical limit of 100% in a few cases.

6.4.8 *2D Configuration*

The initial designs for interfacial evaporation mainly focused on 2D configuration with 2D films or floating substrates creating a localized hot zone at the air–water interface. For this purpose, photothermal materials were embedded in substrates such as membranes, paper, gauze, fabrics, and films to float directly above water.

Membranes: A large number of membranes have been reported as 2D configuration evaporation systems due to their low density, porosity, hydrophilicity, and low cost. Wang *et al.* [2017] combined 2D rGO and 1D multiwalled carbon nanotubes (MWCNTs) and deposited them on a PVDF membrane to obtain a flexible membrane having excellent wettability and a photothermal efficiency of 80.4% under

1 sun illumination. Li *et al.* [2017] vacuum filtered MXene (Ti_3C_2) on PVDF membrane and produced a light-to-heat conversion efficiency of 84% under 1 sun illumination. Tao *et al.* [2018] reported a laminar solar absorber using graphite powder (GP) on a semipermeable collodion membrane (SCM) with highly efficient water evaporation. The GP/SCM composite efficiently converted the absorbed solar energy to heat energy, for example, the 8-mg GP/SCM (i.e., the GP/SCM with 8 mg GP added) enabled water evaporation with high efficiencies of 56.8% under 1.5 kW m^{-2} irradiation and 65.8% under 3 kW m^{-2} irradiation.

Paper: Paper substrates also have similar favorable characteristics as membranes and gauzes for solar-driven steam generation. They are also inexpensive and abundant in nature. Air-laid paper has been utilized the most in this regard owing to its better mechanical strength when compared to traditional paper or tissue paper. The air-laid paper which is prepared from fluff pulp utilizes air instead of water as the medium in the paper-making process. When wetted with water, the paper maintains its strength and shape and thus is an ideal substrate for solar steam generation. Researchers have reported the deposition of rGO, carbon particles, etc., on air-laid paper for 2D substrate–based SISG [Lou *et al.*, 2016; Wang *et al.*, 2017].

Fabrics: Fabrics have added advantages of being extremely flexible, possessing high mechanical strength and durability compared to membranes, gauzes, and paper substrates. Among fabrics, cotton has been used the most owing to its hydrophilicity, low cost, abundance, low thermal conductivity, porosity, and mechanical strength. Cotton fabric has been utilized as a substrate for coating photothermal materials like CB [Liu *et al.*, 2015].

Superhydrophobic cotton fabric was achieved by coating CB and enabling it to float on the water surface for 2D interfacial evaporation (Fig 6.3). Fang *et al.* [2019] used activated carbon fiber cloth (ACFC) with hierarchical microstructures and cotton fiber nonwoven fabrics (CFNF) to bring about a high evaporation rate of 1.59 kg m^{-2}h^{-1} with an optimum conversion efficiency of 93.3% under 1 sun.

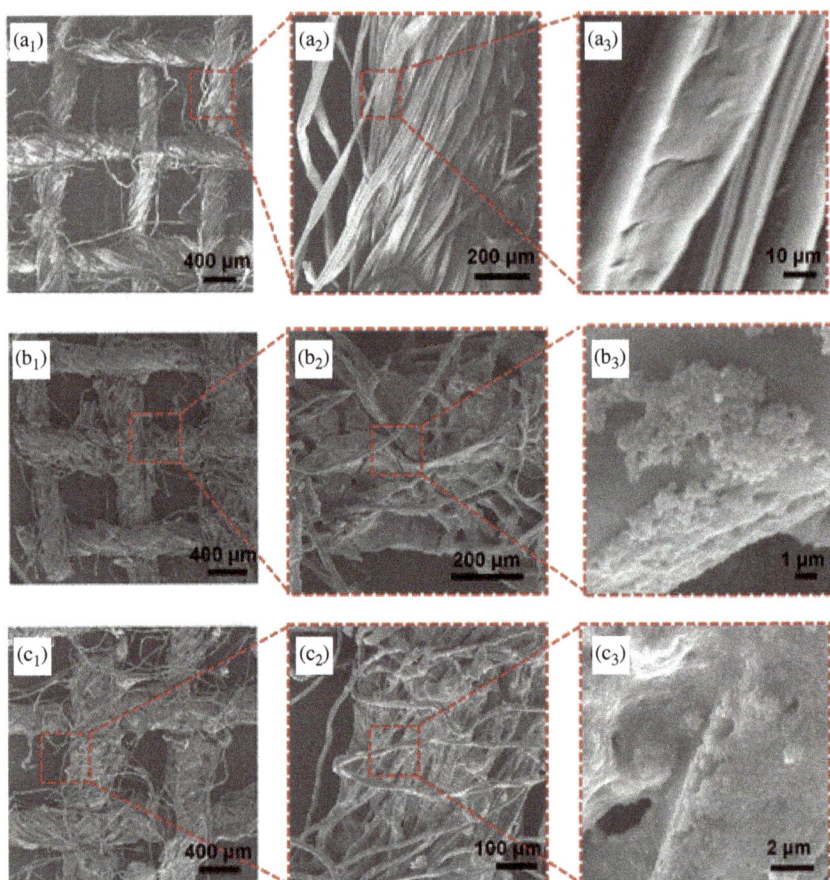

Fig 6.3   SEM images of pristine cotton (a_1-a_3), candle soot-cotton (b_1-b_3), and candle soot-PDMS-cotton gauze (c_1-c_3). Reprinted with permission from Liu *et al.* [2015]. Copyright (2023) American Chemical Society.

Meanwhile, Wilson *et al.* [2019] reported candle-soot and diesel-soot-coated cotton fabrics for excellent solar steam generation. With the help of polystyrene foam as a thermal barrier, the candle-soot-coated and diesel-soot-coated fabrics achieved photothermal efficiencies of 80% and 91.75%, respectively, under 1 sun illumination.

Nylon was utilized to coat candle soot for SISG which achieved a solar thermal conversion efficiency of 58.6% under 1 sun illumination. The silk fabric and rGO (RGO–silk fabric) showed efficient steam generation [Zhang *et al.*, 2018] due to the unique fabric structure of silk and the broadband absorption by rGO, the RGO–silk fabric system with a thermal insulator (polyethylene foam) exhibited remarkably high photothermal performances (an evaporation rate of 1.48 kg m^{-2} h^{-1}) under 1 sun irradiation (1 kW m^{-2}).

Films: Substrate-less photothermal films having excellent photothermal capabilities have also been utilized for 2D solar steam generation, for example, GO film developed as a low-cost, scalable, and porous structure for photothermal conversion [Li *et al.*, 2016]. The GO film was kept upon a polystyrene foam and cellulose fabric setup to achieve 80% solar thermal conversion efficiency under 1 sun illumination. Yang *et al.* [2018] fabricated an ultrathin 2D porous photothermal film of thickness 200 nm based on MoS$_2$ nanosheets and single-walled nanotubes (SWNTs) films. The ultrathin and porous structure effectively facilitated the fast water vapor escape which resulted in an impressively high evaporation efficiency of 91.5%. Liu *et al.* [2019] fabricated sawdust sintered film (SS film) by daubing carbon particles on the sintered sawdust film. They varied the porosity of SS film and found that 0.52 is the optimum porosity for efficient water transportation. Under a solar light power of 1 kW m^{-2}, the optimal porosity gave an evaporation efficiency of 77.64%.

6.4.9 *3D Configuration*

In recent years, 3D configuration-based interfacial evaporators have gained immense interest among researchers for SISG. Owing to its 3D structure for enhanced light absorption, intrinsic water channels and porosity for rapid water transportation, and heat insulation properties for excellent thermal management, 3D configuration evaporation systems have reported extremely high evaporation rates and photothermal efficiencies. 3D configurations reported to date can be categorized primarily as gels, foams, sponges, wood, and so on as described below.

Gels: Aerogels have an excellent 3D porous architecture for solar steam generation. Moreover, aerogels are hydrophilic, super-absorbent, and most importantly ultra-lightweight with extremely low density [Jiang *et al.*, 2018; Hu *et al.*, 2017]. In recent years, graphene-based aerogels have been extensively reported as 3D solar steam generators.

For example, a graphene aerogel (GA) was fabricated from GOs by photoreduction [Fu *et al.*, 2017]. Solar steam generation efficiencies of $53.6 \pm 2.5\%$ and $82.7 \pm 2.5\%$ were achieved at light intensities of 1 and 10 kW m^{-2}, respectively. The authors introduced a new solar absorber of modified graphene aerogel (MGA) prepared by a one-step hydrothermal method and oxygen plasma treatment for rapid and efficient solar steam generation. The obtained MGA possessed interconnected 3D porous networks formed by random construction of rGO sheets and hydrophilic surface due to effective oxygen implant by oxygen plasma (Fig 6.4). Furthermore, under 1 sun illumination, MGA exhibited a high evaporation efficiency of 76.9%.

Hydrogels are another type of gels introduced in 3D solar steam generation. Hydrogels are soft, biocompatible, and extremely hydrophilic materials that can absorb and release water efficiently due to

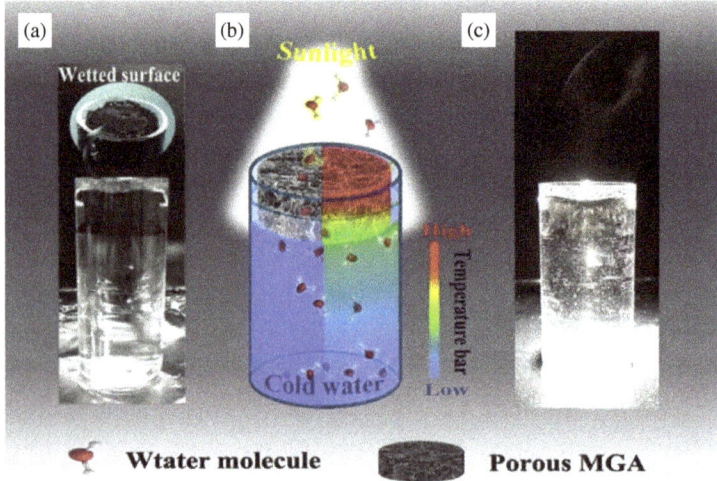

Fig 6.4 (a, c) Optical photographs of porous MGA and (b) schematic representation of solar steam generation in porous MGA. Reprinted with permission from Fu *et al.* [2018]. Copyright (2023) Elsevier.

their macroporous structure. Yin *et al.* [2018] demonstrated a double-network hydrogel with a porous structure (p-PEGDA-PANi) as a flexible, recyclable, and efficient photothermal platform for low-cost and scalable solar steam generation. The hydrogel-based solar steam generator exhibited a maximum solar thermal efficiency of 91.5% with an evaporation rate of 1.40 kg m^{-2}h^{-1} under 1 sun illumination. Guo *et al.* [2020] demonstrated a stabilized Mo_2S_3 by FeS_2 ($Fe-Mo_2S_3$) in sodium alginate hydrogel as a porous solar evaporation system having an excellent evaporation rate of ~2.4 kg m^{-2}h^{-1} with 91.4% solar thermal efficiency under 1 sun illumination.

Wood: Wood is an effective 3D substrate for solar steam generation due to its abundance, low thermal conductivity and density, tremendous mechanical strength, cost-effectiveness, and biocompatibility. In addition, the long cylindrical microchannels present in wood having a high aspect ratio and running parallel to the trunk of the tree provide excellent transport of water from the bulk fluid to the localized heat zone. The absorptive nature of hydrophilic polysaccharides (cellulose microfibrils and hemicellulose) embedded within the lignocellulosic walls maintains a continuous flow of water and the innate mesoporous wood structure induces a capillary effect due to the fiber tracheids and vessels [Zhu *et al.*, 2017]. Various photothermal materials have been reported in literature having wood as a supporting 3D substrate for solar steam generation. GO, PDA, Pd, Ag, Au, CNT, Graphite, $CuFeS_2$, polypyrrole (PPy), and aluminophosphate all have been coated on wood as supporting substrates for efficient steam generation. Liu *et al.* [2018] introduced a novel bilayered structure composed of wood and GO for highly efficient solar steam generation. The wood served as a thermal insulator to confine the photothermal heat to the evaporative surface and to facilitate the efficient transport of water from the bulk to the photothermally active space. Owing to the tailored bilayer structure and the optimal thermo-optical properties of the individual components, the wood–GO composite exhibited a solar thermal efficiency of ~83% under simulated solar excitation at a power density of 12 kW m^{-2}.

Carbonization of wood is another technique extensively reported in 3D wood-based solar steam generation. For example, a solar steam generation device was designed by carbonizing basswood at the top surface to create a bilayer structure with excellent photothermal performance [Zhu *et al.*, 2017]. One-step carbonization technique was also utilized in their comparative study of different carbonized wood-based (poplar, pine, and cocobolo solar steam generators [Jia *et al.*, 2017]. They found out that different wood species show an enormous difference in solar steam generation efficiency due to their porosity. For instance, the energy conversion efficiency of the poplar wood having a porosity of 66% was as high as 86.7% under an intensity of 10 k Wm^{-2} whereas pine wood and cocobolo wood having 60% and 27% porosities had 76.3% and 46.8%, respectively. Xue *et al.* [2017] fabricated carbonized wood using alcohol flame and achieved a solar thermal efficiency of 72% under 1 sun illumination. Liu *et al.* [2018] designed an artificial tree with a reverse tree design for efficient steam generation. In this design, water is transported perpendicular to the direction of natural tree growth, and the top surface is carbonized for light absorption. The artificial tree design achieved a high efficiency of 89% at 10 k W m^{-2}.

Sponges: Like gels and woods, sponges are also low-cost, 3D porous structures having extremely low thermal conductivity and density which makes them an excellent 3D substrate for solar steam generation. Moreover, sponges also have excellent elasticity and flexibility capable of handling external stress effortlessly. In this regard, polyurethane, melamine, polyvinyl alcohol, cellulose, and biomass-derived sponges have been reported to date for efficient 3D steam generation. For instance, Ma *et al.* [2017] reported a recycled black polyurethane sponge coated with polydopamine for SISG and achieved 52.5% under 1 sun illumination. Zhang group introduced a surface-modified polyurethane sponge with bilayered structures for efficient solar steam generation [Zhang *et al.*, 2019]. The top layer was coated with polydimethylsiloxane-modified graphite powder, for light-to-heat conversion and the lower part of the sponge acted as a thermal insulator with a low thermal conductivity in the wet state (0.13882 W m^{-1} K^{-1}).

The bilayered sponge exhibited an evaporation efficiency of 73.3% under 1 sun illumination.

Foams: For 3D solar steam generation, various foams like carbon foam, biofoam, and melamine have been reported as substrates for efficient solar steam generators. For example, authors innovated the use of foam as a 3D structure for solar steam generation with carbon foam [Ghasemi *et al.*, 2014]. Graphite powder on carbon foam was able to effectively localize the heat at the air–water interface and achieve a solar thermal efficiency of 64% and 85% under 1 and 10 sun illuminations. Bilayered hybrid biofoam composed of a BNC layer and an rGO-filled BNC layer exhibited a solar thermal efficiency of 83% under simulated solar illumination (10 kW m^{-2}) [Jiang *et al.*, 2016]. Wu *et al.* [2019] devised plasma-enabled multifunctional all-carbon nanoarchitecture with on-surface waterways formed by nitrogen-doped hydrophilic graphene nanopetals (N-fGPs) seamlessly integrated onto the external surface of hydrophobic self-assembled graphene foam (sGF). The sGF ensured effective thermal insulation and enhanced heat localization, contributing to high solar-vapor efficiency of 88.6 ± 2.1% under 1 sun illumination. Liang *et al.* [2019] reported graphene foam with ~100% solar thermal conversion efficiency under 1 sun illumination. The high solar thermal conversion efficiency was due to the unique injection control technique (ICT) adapted to form a capillary water-like state in rGO foam solar absorber. The ICT technique controls the rate of water supply, using the equilibrium relationship between water supply rate and evaporation rate. This creates an incomplete absorption of water by rGO foams which helps in keeping the pores open and increases the contact area between air and water promoting efficient energy utilization.

Melamine foams (MFs), owing to their porous structure, thermal insulation properties, and lightweight, have been utilized extensively for 3D solar steam generation, for example, carbonized MFs (ISA) through one-step calcination, for water evaporation [Lin *et al.*, 2018]. The optimized ISA exhibited a water evaporation rate of 1.270 kg m^{-2} h^{-1} and an energy conversion efficiency of 87.3% under

1 kWm^{-2} solar illumination. Authors pre-pressed the MF for efficient and stable solar steam generation [Li *et al.*, 2019]. Making use of the bilayer structure, different functions were assigned to different layers, with Ppy coating pre-pressed MF (PPy layer) for light absorption and water evaporation, and bottom pre-pressed MF layer for water transport and thermal insulation. A high average evaporation rate of 1.574 kg m^{-2} h^{-1} and a superb steam generation efficiency of 90.4% under 1 sun were achieved.

Carbonized structures: Structures such as mushrooms, lotus seedpods, rice straw, magnolia fruit, sunflower head, corn straw, and sugarcane having unique physical attributes have been also reported for nature-derived 3D solar steam generation. For example, carbonized mushrooms achieved 78% conversion efficiency under 1 sun illumination. The capability of high solar steam generation was attributed to the unique natural structure of mushrooms, umbrella-shaped black pileus, porous context, and fibrous stipe with a small cross-section. These features not only provided efficient light absorption, water supply, and vapor escape, but also suppressed different heat losses (convection, conduction, and radiation) at the same time. Fang *et al.* carbonized lotus seedpods to achieve an evaporation rate and the corresponding evaporation efficiency under 1 sun irradiation was 1.30 kg m^{-2} h^{-1} and 86.5%, respectively. The excellent performance of carbonized lotus seedpods was attributed to the unique macroscopic cone shape and hierarchical meso/macropore structures forming an interconnected porous network.

Wang *et al.* [2018] developed a 3D photothermal cone (Ppy-paper) for high-efficiency solar-driven evaporation with minimum light reflection and heat loss to bulk water. The 3D cone could absorb almost 99.2% incoming radiation and provide excellent heat localization due to its conical design. A solar conversion efficiency of 93.8% for evaporation was achieved by the cone design under 1 sun illumination. Few researchers demonstrated water boiling and steam generation under nonconcentrated ambient solar flux in a receiver open to the ambient. The lab-scale 1 sun, ambient steam generator (OAS) had three components. One, a spectrally selective solar absorber made of a cermet (Bluetec eta plus) coated on a copper sheet; second, a thermal insulator using polystyrene foam disk; and third, a convective cover made from a sheet of large transparent bubble wrap. OAS achieved excellent thermal insulation due to polystyrene foam and bubble wrap which obtained a photothermal efficiency of <80% under 1 sun illumination.

6.5 Applications of Interfacial Systems

The development of configurations with high solar thermal efficiencies has brought great attention to the application of interfacial systems for practical purposes such as desalination, water purification, dye degradation, distillation, electricity production, sterilization, and so on.

Desalination: Seawater desalination utilizing solar energy has gained tremendous interest in recent years due to the growing development of extremely efficient interfacial systems. A simple approach followed is the integration of interfacial systems replacing black paint in solar stills for better solar thermal conversion and steam generation. A typical solar-derived clean water generation process can be detailed in three steps: (1) efficient solar absorption and heat generation; (2) high water vaporization performance due to generated heat; and (3) effective vapor condensation and collection. The performance of solar thermal devices can be quantified with respect to their specific water productivity (SWP) [Wang *et al.*, 2019]. SWP can be defined as the volume of water generated per area per time. SWP considers all the three steps of clean water generation as shown in the equation.

$$\mathrm{SWP} = \frac{I}{h} \times \propto \times \eta \times \mathrm{GOR} \qquad (6.2)$$

where I is the solar irradiance (kW m^{-2}), h is the latent heat of vaporization, $\propto$ is the absorptivity, η is the solar thermal efficiency, and GOR is the ratio of output clean water produced (kg) per vapor generated (kg). Hence, an efficient evaporation system such as SISG plays a significant role in solar still–based large-scale clean water generation owing to the high solar absorptivity and solar thermal efficiencies of 2D or 3D solar steam generators.

Since the introduction of 2D and 3D structured interfacial evaporators, the utilization of these devices for highly efficient desalination has skyrocketed. In a typical desalination process, seawater is evaporated due to photothermal effects of steam generators. This evaporation leads to the separation of salt from seawater and generates clean water which can be collected by condensation. Almost all of the desalination experiments have reported excellent salt removal rates (~100%) with salt concentrations well below the permitted WHO guidelines for drinking water [Wilson *et al.*, 2020].

However, the application of seawater desalination in large-scale, long-term applications can be hampered by its low SWP. This is due to the poor condensation of evaporated vapors which significantly affects the GOR of the solar still. In traditional passive solar still, the condensation of evaporated vapors is extremely inefficient due to the poor housing design of the solar still. Salt accumulation is another aspect of interfacial evaporation that could adversely affect the steam generation performance in long-term applications. Hence, several studies have reported the development of salt-resistant or anti-clogging structures for desalination. CB cotton gauze was demonstrated to develop a self-cleaning ability in which NaCl crystals were automatically rejected by the superhydrophobicity of the membrane [Liu *et al.*, 2015].

Even though interfacial evaporators have improved steam generation significantly, considerable development is still required for the stable and reusable applicability of these evaporators in long-term operations. Moreover, the issues such as operational costs, washing of solar stills, housing design, and salt accumulation needs to be addressed in detail in solar-driven interfacial evaporation-based desalination for practical long-term operations.

Water purification: Similar to the seawater desalination mechanism, the evaporation of water molecules through solar energy produces a pure distillate that is completely free of pollutants. Even though interfacial evaporation promotes water vaporization well below the boiling point of water, the separation of water molecules from contaminants is not affected and the contaminants are left behind as a residue in the water reservoir. This mechanism has thus prompted many researchers to utilize solar-driven interfacial evaporation for all kinds of water contaminant removal. For instance, Wang *et al.* [2019] reported a high evaporation performance of water samples containing sanitary wastewater, 0.1M H_2SO_4 and 0.1M NaOH indicating application of Ppy-coated membranes for clean water production. Wilson *et al.* [2020] reported the purification of sewage to obtain clean water using a floating candle-soot-coated cotton solar still (Fig 6.5). Moreover, a 90% removal of total organic carbon from industrial

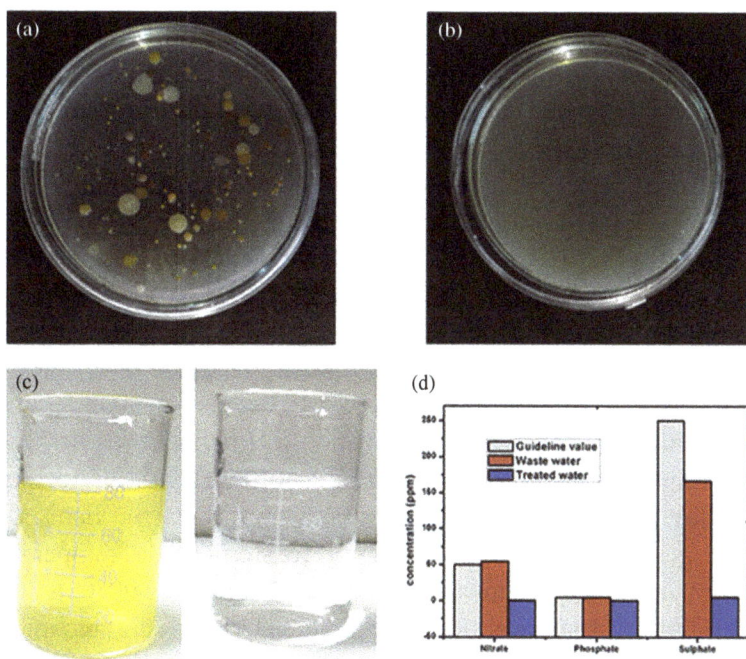

Fig 6.5 (a) Digital photographs of bacteria colonies in sewage, (b) treated water, (c) Digital photographs of sewage (yellowish) and treated water respectively and (d) The bar graph illustrating the nitrate, sulfate, and phosphate values of sewage and treated water. Reprinted with permission from Wilson *et al.* [2019]. Copyright (2023) Elsevier.

wastewater was also demonstrated using a facile carbon fabric-based prototype [Wilson *et al.*, 2018].

Dye removal: The removal of toxic dye molecules, such as RhB, which are frequently found in industrial wastewater, using solar energy has been explored greatly in recent years. Two approaches have been mainly followed in this regard: (1) photocatalytic dye degradation and (2) evaporation–condensation of dye solution. Photocatalytic dye degradation using interfacial technique has been reported extensively due to its unique advantages over particle-based systems such as large surface area, enhanced light absorption, and reusability. For instance, TiO_2 has been reported as a photocatalytic material in TiO_2-Au-AAO membrane to create a bifunctional membrane capable of purifying contaminated water containing RhB solution [Liu *et al.*, 2016]. Xe-light illumination was able to degrade 60% of dye contamination under 2 h yielding clean water through evaporation–condensation mechanism. Meanwhile, Deng *et al.* [2019] studied the combination of dyes (methylene blue, Rhodamine B, and methyl orange) using evaporation–condensation mechanism to obtain clean water with no dye present in them. Authors examined the generation of clean water using Rhodamine B, methylene blue, and methyl orange as model pollutants [Guo *et al.*, 2020]. The collected water contained negligible amounts of pollutants after evaporation–condensation mechanism.

Sterilization: A typical vapor sterilization technique in medical applications requires a minimum heating of vapor to 121°C for 15 min or 132°C for 5 min for fool-proof sterilization. This mechanism has three stages, namely heating of vapor to sterilization temperature, maintaining the temperature, and finally cooling to 100°C to finish the sterilization. In conventional, commercial sterilization techniques, bulk heating of water is followed which is time-consuming and energy inefficient with the requirement of electricity. Researchers have then reported the elimination of these problems using SISG. Zhang *et al.* [2017] demonstrated successful sterilization using rGO/PTFE-composite membrane which was capable of producing steam at temperatures higher than 120°C. The sterilization device was able to demonstrate sterilization capacity in both chemical and biological sterilization tests.

Electricity generation: Electricity generation by photovoltaics has been prevalent in the renewable energy industry for the past few decades. It highly depends on the bandgap of the photovoltaic semi-conductor material to generate electron–hole pairs. However, photon energies lower than the bandgap energy of the semiconductor do not initiate electron–hole generation and are thus wasted as heat energy. Researchers have now focused on interfacial membrane techniques to generate electricity capable of absorbing the entire solar spectrum. For instance, Yang *et al.* [2017] utilized conducting CNT and Nafion membrane to simultaneously generate steam and electricity. The hybrid device was tested for different solar intensities and stable output powers of 0.5 W m^{-2} and 1.1 W m^{-2} were obtained under 1 and 2 sun illuminations, respectively. The device was able to generate a stable output power of 0.5 W m^{-2} for 20 h.

6.6 Conclusion and Perspective

Solar-driven steam generation has gained considerable momentum in recent years with broadband solar absorption and extremely high photothermal efficiencies. In this regard, significant progress has been made in the development of solar steam generators with photo-thermal materials utilized for solar thermal conversion, design struc-tures optimized for heat localization, thermal insulation, and water path techniques, and finally the applications in seawater desalination, wastewater purification, sterilization, and so on. Despite the develop-ment of extremely efficient solar steam generators, the practical appli-cations of these generators remain underdeveloped for large-scale and long-term applications. Moreover, various challenges and issues still remain that need to be addressed before large-scale applications. First, the correct evaluation of solar thermal efficiency remains con-flicted with different reports having varying test conditions such as room temperature, wind conditions, and humidity, which affects the evaporation performance of the system. Second, the perfect balance of low cost, scalability, feasibility of fabrication, and high evaporation performance is still in progress even after reports of photothermal efficiencies over 100% under 1 sun illumination. Third, the application of interfacial systems in solar seawater desalination is still in its initial stages with very low SWP and salt accumulation issues hindering its usage for practical purposes. In addition, most of the applications are in laboratory stages and require considerable progress for practical applications. Solar-driven steam generation has shown considerable development in the past few years with efficiencies more than 100% achieved under 1 sun illuminations. In the future, we can expect considerable development of solar steam generators in practical applications capable of replacing existing technologies in a cost-effective and energy-efficient manner.

References

Abdelrahman, M., Fumeaux, P. & Suter, P. (1979). Study of solid-gas-suspensions used for direct absorption of concentrated solar radiation. *Sol. Energy* 22(1), pp. 45–48.

Anderson, M. A., Cudero, A. L. & Palma, J. (2010). Capacitive deionization as an electrochemical means of saving energy and delivering clean water. Comparison to present desalination practices: Will it compete? *Electrochim. Acta* 55(12), pp. 3845–3856.

Deng, Z., Zhou, J., Miao, L., Liu, C., Peng, Y., Sun, L. & Tanemura, S. (2017). The emergence of solar thermal utilization: Solar-driven steam generation. *J. Mater. Chem. A Mater. Energy Sustain.* 5(17), pp. 7691–7709.

Fu, Y., Wang, G., Mei, T., Li, J., Wang, J. & Wang, X. (2017). Accessible graphene aerogel for efficiently harvesting solar energy. *ACS Sustain. Chem. Eng.* 5(6), pp. 4665–4671.

Fu, Y., Wang, G., Ming, X., Liu, X., Hou, B., Mei, T. & Wang, X. (2018). Oxygen plasma treated graphene aerogel as a solar absorber for rapid and efficient solar steam generation. *Carbon* 130, pp. 250–256.

Gao, M., Zhu, L., Peh, C. K. & Ho, G. W. (2019). Solar absorber material and system designs for photothermal water vaporization towards clean water and energy production. *Energy Environ. Sci.* 12(3), pp. 841–864.

Ghafurian, M. M., Niazmand, H. & Ebrahiminia-Bajestan, E. (2019). Improving steam generation and distilled water production by volumetric solar heating. *Appl. Therm. Eng.* 158(113808), p. 113808.

Ghasemi, H., Ni, G., Marconnet, A. M., Loomis, J., Yerci, S., Miljkovic, N. & Chen, G. (2014). Solar steam generation by heat localization. *Nat. Commun.* 5(1), p. 4449.

Ghim, D., Jiang, Q., Cao, S., Singamaneni, S. & Jun, Y.-S. (2018). Mechanically interlocked 1T/2H phases of MoS$_2$ nanosheets for solar thermal water purification. *Nano Energy* 53, pp. 949–957.

Gueymard, C. A. (2004). The sun's total and spectral irradiance for solar energy applications and solar radiation models. *Sol. Energy* 76(4), pp. 423–453.

Guo, Z., Yu, F., Chen, Z., Shi, Z., Wang, J. & Wang, X. (2020). Stabilized Mo2S3 by FeS2 based porous solar evaporation systems for highly efficient clean freshwater collection. *Sol. Energy Mater. Sol. Cells* 211(110531), p. 110531.

Hou, B., Cui, Z., Zhu, X., Liu, X., Wang, G., Wang, J. & Wang, X. (2019). Functionalized carbon materials for efficient solar steam and electricity generation. *Mater. Chem. Phys.* 222, pp. 159–164.

Hu, X., Xu, W., Zhou, L., Tan, Y., Wang, Y., Zhu, S. & Zhu, J. (2017). Tailoring graphene oxide-based aerogels for efficient solar steam generation under one sun. *Adv. Mater.* 29(5), p. 1604031.

Jia, C., Li, Y., Yang, Z., Chen, G., Yao, Y., Jiang, F., Kuang, Y., Pastel, G., Xie, H., Yang, B., Das, D. & Hu, L. (2017). Rich mesostructures derived from Natural Woods for solar steam generation. *Joule* 1(3), pp. 588–599.

Jiang, F., Liu, H., Li, Y., Kuang, Y., Xu, X., Chen, C. & Hu, L. (2018). Lightweight, mesoporous, and highly absorptive all-nanofiber aerogel for efficient solar steam generation. *ACS Appl. Mater. Interfaces* 10(1), pp. 1104–1112.

Jiang, Q., Tian, L., Liu, K.-K., Tadepalli, S., Raliya, R., Biswas, P. & Singamaneni, S. (2016). Bilayered biofoam for highly efficient solar steam generation. *Adv. Mater.* 28(42), pp. 9400–9407.

Jin, Y., Chang, J., Shi, Y., Shi, L., Hong, S. & Wang, P. (2018). A highly flexible and washable nonwoven photothermal cloth for efficient and practical solar steam generation. *J. Mater. Chem. A Mater. Energy Sustain.* 6(17), pp. 7942–7949.

Li, C., Jiang, D., Huo, B., Ding, M., Huang, C., Jia, D. & Liu, J. (2019). Scalable and robust bilayer polymer foams for highly efficient and stable solar desalination. *Nano Energy* 60, pp. 841–849.

Li, X., Xu, W., Tang, M., Zhou, L., Zhu, B., Zhu, S. & Zhu, J. (2016). Graphene oxide-based efficient and scalable solar desalination under one sun with a confined 2D water path. *Proc. Natl. Acad. Sci. USA* 113(49), pp. 13953–13958.

Liang, H., Liao, Q., Chen, N., Liang, Y., Lv, G., Zhang, P. & Qu, L. (2019). Thermal efficiency of solar steam generation approaching 100% through capillary water transport. *Angew. Chem. Weinheim Bergstr. Ger.* 131(52), pp. 19217–19222.

Lin, X., Chen, J., Yuan, Z., Yang, M., Chen, G., Yu, D. & Chen, X. (2018). Integrative solar absorbers for highly efficient solar steam generation. *J. Mater. Chem. A Mater. Energy Sustain.* 6(11), pp. 4642–4648.

Liu, H., Chen, C., Chen, G., Kuang, Y., Zhao, X., Song, J. & Hu, L. (2018a). High-performance solar steam device with layered channels: Artificial tree with a reversed design. *Adv. Energy Mater.* 8(8), p. 1701616.

Liu, Y., Chen, J., Guo, D., Cao, M. & Jiang, L. (2015). Floatable, self-cleaning, and carbon-black-based superhydrophobic gauze for the solar evaporation enhancement at the air-water interface. *ACS Appl. Mater. Interfaces* 7(24), pp. 13645–13652.

Liu, H., Chen, C., Wen, H., Guo, R., Williams, N. A., Wang, B. & Hu, L. (2018b). Narrow bandgap semiconductor decorated wood membrane for high-efficiency solar-assisted water purification. *J. Mater. Chem. A Mater. Energy Sustain.* 6(39), pp. 18839–18846.

Liu, S., Huang, C., Luo, X. & Guo, C. (2019). Performance optimization of bi-layer solar steam generation system through tuning porosity of bottom layer. *Appl. Energy* 239, pp. 504–513.

Liu, Y., Zhao, J., Zhang, S., Li, D., Zhang, X., Zhao, Q. & Xing, B. (2022). Advances and challenges of broadband solar absorbers for efficient solar steam generation. *Environ. Sci. Nano* 9(7), pp. 2264–2296.

Lou, J., Liu, Y., Wang, Z., Zhao, D., Song, C., Wu, J. & Deng, T. (2016). Bioinspired multifunctional paper-based rGO composites for solar-driven clean water generation. *ACS Appl. Mater. Interfaces* 8(23), pp. 14628–14636.

Ma, S., Chiu, C. P., Zhu, Y., Tang, C. Y., Long, H., Qarony, W. & Tsang, Y. H. (2017). Recycled waste black polyurethane sponges for solar vapor generation and distillation. *Appl. Energy* 206, pp. 63–69.

Neumann, O., Urban, A. S., Day, J., Lal, S., Nordlander, P. & Halas, N. J. (2013). Solar vapor generation enabled by nanoparticles. *ACS Nano* 7(1), pp. 42–49.

Ni, G., Miljkovic, N., Ghasemi, H., Huang, X., Boriskina, S. V., Lin, C.-T. & Chen, G. (2015). Volumetric solar heating of nanofluids for direct vapor generation. *Nano Energy* 17, pp. 290–301.

Ren, H., Tang, M., Guan, B., Wang, K., Yang, J., Wang, F. & Liu, Z. (2017). Hierarchical graphene foam for efficient omnidirectional solar-thermal energy conversion. *Adv. Mater.* 29(38), p. 1702590.

Tahir, Z., Kim, S., Ullah, F., Lee, S., Lee, J.-H., Park, N.-W. & Kim, Y. S. (2020). Highly efficient solar steam generation by glassy carbon foam coated with two-dimensional metal chalcogenides. *ACS Appl. Mater. Interfaces* 12(2), pp. 2490–2496.

Tao, F., Zhang, Y., Wang, B., Zhang, F., Chang, X., Fan, R. & Yin, Y. (2018). Graphite powder/semipermeable collodion membrane composite for water evaporation. *Sol. Energy Mater. Sol. Cells* 180, pp. 34–45.

Tsai, M.-L., Su, S.-H., Chang, J.-K., Tsai, D.-S., Chen, C.-H., Wu, C.-I. & He, J.-H. (2014). Monolayer MoS_2 heterojunction solar cells. *ACS Nano* 8(8), pp. 8317–8322.

Ulset, E. T., Kosinski, P. & Balakin, B. V. (2018). Solar steam in an aqueous carbon black nanofluid. *Appl. Therm. Eng.* 137, pp. 62–65.

Wang, Q., Guo, Q., Jia, F., Li, Y. & Song, S. (2020). Facile preparation of three-dimensional MoS_2 aerogels for highly efficient solar desalination. *ACS Appl. Mater. Interfaces* 12(29), pp. 32673–32680.

Wang, X., He, Y., Cheng, G., Shi, L., Liu, X. & Zhu, J. (2016a). Direct vapor generation through localized solar heating via carbon-nanotube nanofluid. *Energy Convers. Manag.* 130, pp. 176–183.

Wang, Z., Horseman, T., Straub, A. P., Yip, N. Y., Li, D., Elimelech, M. & Lin, S. (2019). Pathways and challenges for efficient solar-thermal desalination. *Sci. Adv.* 5(7), p. 0763.

Wang, Y., Wang, C., Song, X., Huang, M., Megarajan, S. K., Shaukat, S. F. & Jiang, H. (2018). Improved light-harvesting and thermal management for efficient solar-driven water evaporation using 3D photothermal cones. *J. Mater. Chem. A Mater. Energy Sustain.* 6(21), pp. 9874–9881.

Wang, Z., Ye, Q., Liang, X., Xu, J., Chang, C., Song, C. & Deng, T. (2017). Paper-based membranes on silicone floaters for efficient and fast solar-driven interfacial evaporation under one sun. *J. Mater. Chem. A Mater. Energy Sustain.* 5(31), pp. 16359–16368.

Wang, Y., Zhang, L. & Wang, P. (2016b). Self-floating carbon nanotube membrane on macroporous silica substrate for highly efficient solar-driven interfacial water evaporation. *ACS Sustain. Chem. Eng.* 4(3), pp. 1223–1230.

Wilson, H. M., Tushar, R. A. S. & Jha, N. (2020). Plant-derived carbon nanospheres for high efficiency solar-driven steam generation and seawater desalination at low solar intensities. *Sol. Energy Mater. Sol. Cells* 210(110489), p. 110489.

Wu, S., Xiong, G., Yang, H., Gong, B., Tian, Y., Xu, C. & Ostrikov, K. (ken). (2019). Multifunctional solar waterways: Plasma-enabled self-cleaning nanoarchitectures for energy-efficient desalination. *Adv. Energy Mater.* 9(30), p. 1901286.

Xue, G., Liu, K., Chen, Q., Yang, P., Li, J., Ding, T. & Zhou, J. (2017). Robust and low-cost flame-treated wood for high-performance solar steam generation. *ACS Appl. Mater. Interfaces* 9(17), pp. 15052–15057.

Yang, X., Yang, Y., Fu, L., Zou, M., Li, Z., Cao, A. & Yuan, Q. (2018). An ultrathin flexible 2D membrane based on single-walled nanotube-MoS$_2$ hybrid film for high-performance solar steam generation. *Adv. Funct. Mater.* 28(3), pp. 1704505.

Yin, X., Zhang, Y., Guo, Q., Cai, X., Xiao, J., Ding, Z. & Yang, J. (2018). Macroporous double-network hydrogel for high-efficiency solar steam generation under 1 sun illumination. *ACS Appl. Mater. Interfaces* 10(13), pp. 10998–11007.

Zhang, X., Gao, W., Su, X., Wang, F., Liu, B., Wang, J.-J. & Sang, Y. (2018a). Conversion of solar power to chemical energy based on carbon nanoparticle modified photo-thermoelectric generator and electrochemical water splitting system. *Nano Energy* 48, pp. 481–488.

Zhang, Z., Mu, P., He, J., Zhu, Z., Sun, H., Wei, H. & Li, A. (2019). Facile and scalable fabrication of surface-modified sponge for efficient solar steam generation. *Chem. Sus. Chem.* 12(2), pp. 426–433.

Zhang, Q., Xiao, X., Wang, G., Ming, X., Liu, X., Wang, H. & Wang, X. (2018b). Silk-based systems for highly efficient photothermal conversion under one sun: Portability, flexibility, and durability. *J. Mater. Chem. A Mater. Energy Sustain.* 6(35), pp. 17212–17219.

Zhu, M., Li, Y., Chen, G., Jiang, F., Yang, Z., Luo, X. & Hu, L. (2017). Tree-inspired design for high-efficiency water extraction. *Adv. Mater.* 29(44), p. 1704107.

Chapter 7

An Overview of the Recent Developments in Graphene-MoS$_2$-Based Supercapacitors

M. Manuraj[1,*], Visakh V Mohan[2,*], and R. B. Rakhi[3]

[1]*Chemical Sciences and Technology Division, CSIR — National Institute of Interdisciplinary Science and Technology (CSIR-NIIST), Thiruvananthapuram, Kerala 695019, India.*
[2]*Department of Physics, University of Kerala, Kariavattom, Thiruvananthapuram, Kerala 695581, India.*
[3]*The Centre for Sustainable Energy Technologies, CSIR- National Institute of Interdisciplinary Science and Technology (CSIR-NIIST), Thiruvananthapuram, Kerala 695019, India.*
Authors contributed equally
Emails: rakhisarath@gmail.com; rakhiraghavanbaby@niist.res.in

Abstract

The fascinating electrochemical properties of graphene and its derivatives have undoubtedly changed the scope of the supercapacitor field. Since 2004, graphene has been considered an indispensable material in the field of science and technology. Its alluring properties, such as flexibility, lightweightness, mechanical stability, large surface area, high thermal stability, excellent electrical conductivity, and mechanical strength made it a very promising material among various 2D nanomaterials. Supercapacitor electrodes with excellent electrochemical performance can be obtained by combining graphene with various metal oxides, polymer composites, and transition metal dichalcogenides (TMDs). Numerous studies have been conducted to study the combination of graphene with metal oxides and polymer nanocomposites. Currently, much research is focused on supercapacitor application of graphene–TMD combination, owing to the fascinating properties of this combination. MoS_2 is one of the 2D TMD materials that has superior properties compared with other TMDs. Herein, we discuss about the recent advances in the field of supercapacitors based on graphene–MoS_2 combination.

7.1 Introduction

Energy resources can be broadly classified as renewable and nonrenewable. In the past, we had an abundant amount of nonrenewable resources, such as gas, oil, and coal, available on earth to satisfy our energy needs. The increase in human population and the development of technology have contributed to a tremendous increase in global energy requirements, leading to the depletion of the available conventional energy resources. Hence, modern society is forced to think about alternative and renewable energy resources such as solar, wind, and tide. However, the problem with these renewable resources is their intermittency. The sun will not shine during night, and we cannot produce enough power from solar radiation during the rainy season. Similarly, the wind is not going to blow whenever we need it. That means we cannot use all these energy sources continuously. For satisfying our energy needs, the energy produced from renewable resources needs to be stored, and here lies the significance of energy storage devices in modern life.

The widespread use of portable electronic devices in our day-to-day life has led to the development of various energy storage devices having high storage capacity, long charge–discharge life, and short charging time. We have plenty of energy storage devices for applications, but the choice and performance demands of these devices are still challenging [Winter & Brodd, 2004]. The next candidates are batteries that can store more charges via faradaic redox reactions. The technological revolution in the area of storage devices began with the commercialization of Lithium ions by Sony in 1991 and within no time, it became an integral part of many portable devices and also found its usage in the field of transportation. The requirements in the field of portable electronics as well as electronic vehicles are changing day by day and their performance demands are also increasing in multiple ways with fast charging being a major requirement. Compared to any other storage device, batteries can store more electrical energy. However, their slow ionic diffusion causes long charging time (1–10 h). Compared to these energy storage devices, supercapacitors are recognized as promising storage devices having applications, particularly in hybrid vehicles. Their importance is due to their high energy density than conventional dielectric capacitors with attractive properties like fast charging–discharging, long cycle life, long shelf life, and a broad range of operating temperatures.

7.2 Dielectric Capacitor and Supercapacitor

A dielectric capacitor, also known as an electrostatic capacitor, consists of two metal plates that are conductive in nature and are usually made up of aluminum, tantalum, or other metals separated by a dielectric material like paper, glass, ceramics, or any other materials which obstructs the flow of electrons. The charging mechanism of the dielectric capacitor is simple. When these metal plates are connected to a power source, it induces a potential as the metal plates contain a large number of positive and negative charges. The dielectric material, being an insulator, blocks the flow of electrons. Thus, the charge will accumulate in both metal plates based on the polarity of the external source. During the discharge time, the capacitor releases the energy through the external load connected to it. But the applications of these conventional capacitors are limited to low-energy applications. Hence, for the high energy applications, researchers started looking out for new materials, which led to the fabrication of supercapacitors. The major difference between a normal capacitor and an ESs is the electrode material. The ES electrodes are composed of highly porous materials like nanocarbons and consist of solid or liquid electrolytes in the place of dielectric, so that in electrode/electrolyte interface, at which the charge accumulation occurs, forms the electric double layer. Due to the large surface area, ES can store more electric charges than normal capacitors, and also by modifying the electrode surface we can further improve the capacitance of ESs.

7.3 History of Supercapacitors

The double-layer formation at the electrode/electrolyte interface, which is the fundamental phenomenon that occurs in supercapacitors, was first predicted by Helmholtz in 1879. However, the practical use of these double-layer capacitors was demonstrated only after the 1950s. The first electric double-layer capacitor based on activated carbon was developed by H. I Becker of General Electric (GE) and patented this development. However, GE did not follow this development. Later, the first electric double-layer capacitor for commercial application was developed by Becker Standard Oil Company of OHIO, Japan, in 1969 using activated carbon and tetra alkyl ammonium salt in a nonaqueous solvent as an electrolyte. In 1978, the Nippon Electric Company (NEC) Japan marketed the double-layer capacitor for computer backup memory. The low, internal-resistance supercapacitor for military applications was developed by Pinnacle Research Institute (PRI) in 1982, and subsequently, Maxwell Technologies took over the research of PRI on supercapacitors and marketed the supercapacitor in the name of "BOOST CAPS." Although marketed, the supercapacitors' specific energy values were still challenging up to the 90s, with no significant developments from the materials' point of view. Later, in 1999, B E Conway introduced a new storage mechanism known as pseudocapacitance with RuO_2 as the electrode material. The research interest in supercapacitors was piqued up after recognizing the importance of hybrid vehicles leading to the development of new pseudocapacitive materials which give 10–100 times higher energy density than the double-layer capacitor.

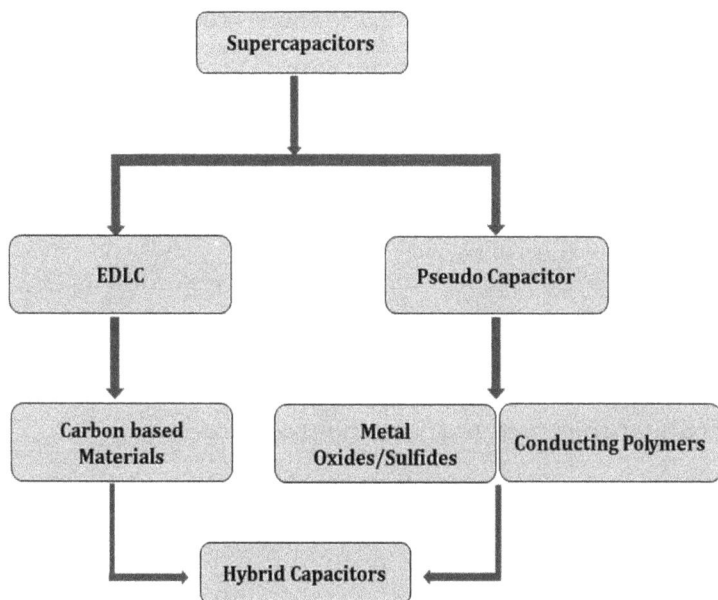

Fig 7.1   Classification of supercapacitors based on the storage mechanism.

7.4 Taxonomy of Supercapacitors

Depending on the functional mechanisms, supercapacitors can be classified into three: electric double-layer capacitors, pseudocapacitors, and hybrid capacitors (Fig 7.1). All these supercapacitors are classified based on their storage mechanisms like nonfaradaic, faradaic redox reaction, or a combination of these two. In the faradaic process, the charge storage mechanism involves oxidation and reduction of the electrode materials whereas in the nonfaradaic process, it is through static charges. This means that in the faradaic process there is a chemical reaction involved, whereas in the nonfaradaic process no such chemical reaction is present, it is a purely physical process [Huang *et al.*, 2014].

7.4.1 *Electric Double-Layer Capacitor (EDLCs)*

EDLCs are constructed using two electrodes made up of carbon-based materials, an electrolyte, and a separator. Like conventional dielectric capacitors, double-layer capacitors also store energy using static charges which, being a nonfaradaic process in nature, have no charge transfer between the electrode and the electrolyte. These types of supercapacitors use the double-layer mechanism for the storage of charges. When a voltage is applied across the electrodes, the charges get accumulated on the electrode surface, and oppositely charged ions in the electrolyte diffuse through the separator and move toward the electrode surface so that double layers are formed on both of the electrodes which are similar to the series connection of two parallel plate capacitors. These double layers will increase the energy density by increasing the surface area and decreasing the separation between the electrodes compared to conventional capacitors (Fig 7.2).

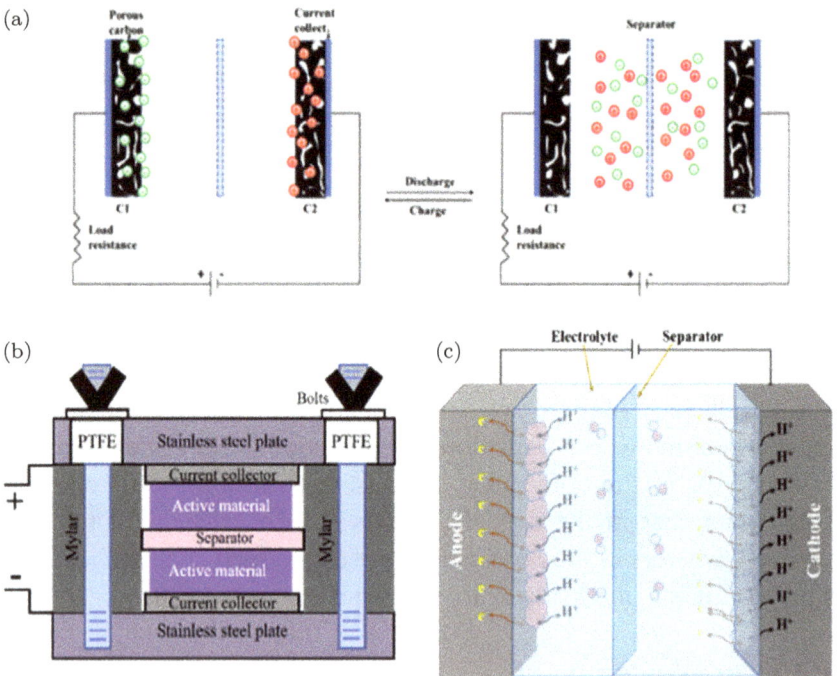

Fig 7.2 Schematic of (a) charging and discharging of the supercapacitor, (b) pseusocapacitor, and (c) two-electrode configuration.

No chemical reaction occurs at the electrode surface as there is no charge transfer between the electrode and electrolyte. This will improve the cyclic stability of the capacitors. Generally, the EDLCs show high cyclic stability around one million charge–discharge cycles compared to other electrochemical batteries which show only 1000 cycles. Because of this high stability, EDLCs can be used for some deep-sea applications and other nonuser environmental applications. The performance of EDLC can also be tuned by changing the types of electrolytes that are used. Generally, aqueous-based electrolytes are used for EDLCs which have low equivalent series resistance (ESR) values, but the breakdown voltage of aqueous electrolytes is lower compared to organic electrolytes, so the selection of electrolytes depends on the applications in which EDLCs are used.

7.4.2 *Pseudocapacitors*

Pseudocapacitors store energy through surface faradaic redox reactions between the electrode and electrolyte. These faradaic redox reactions allow pseudocapacitors to exhibit higher capacitance and energy density compared to EDLCs. Transition metal oxides and conducting polymers are mostly used as electrode materials for pseudocapacitors. As the charge storage mechanism is mainly a redox reaction, these supercapacitors are somewhat behaving like batteries, and, their stability is less compared to EDLCs [Conway, 1991].

7.4.3 *Hybrid Capacitor*

To exploit the advantages of both EDLCs and pseudocapacitor materials and to mitigate the disadvantages of these two, scientists started attempting to develop hybrid capacitors. Charging the capacitor by utilizing both faradaic and nonfaradaic processes, hybrid capacitors have achieved both high energy density and power density without sacrificing the stability of the capacitor. There are different types of hybrid capacitors such as hybrid composite supercapacitors and asymmetric capacitors [Jiang & Liu *et al.*, 2019].

7.4.4 *Composite Electrodes*

Composite electrodes integrate carbon materials with other conducting polymers or metal oxides by physical or chemical methods. Here the carbon materials will act as a support, which facilitates double-layer capacitance and provide high active surface area for electrolyte ions. The use of carbon materials will increase the stability and the pseudocapacitor materials can improve the capacitance value. The use of graphene and carbon nanotubes with metal oxides like MnO_2, RuO_2, NiO and conducting polymers like polyaniline, polypyrrole have been well studied and it has been demonstrated that their combinations can achieve better capacitive performance [Ren *et al.*, 2018].

7.4.5 *Asymmetric Capacitor*

Asymmetric hybrids combine faradaic and nonfaradaic mechanisms by incorporating an EDLC electrode with a pseudocapacitive electrode. Carbon-based materials are used as the negative electrode and metal oxides or other conducting polymers as the positive electrode. This construction may increase the voltage window of the capacitor thereby leading to higher energy density and power density than EDLCs [Tiwari *et al.*, 2021].

7.5 Supercapacitor Designing for Testing

7.5.1 *Two-Electrode Configuration*

In the two-electrode system, the same active material is deposited over two current collectors to fabricate the cathode and anode. The electrodes are deposited over the current collectors through drop casting, spin coating, printing, or any other deposition techniques. A separator is placed in between these two electrodes to avoid a short circuit. The assembly is (two electrodes and a separator) dipped into a suitable electrolyte depending on the application and electrode materials. The capacitor is assembled into a metal case which can be tightened by using screws without any outside contact.

7.5.2 *Electrochemical Analysis*

Capacitance is the key parameter to measure the amount of energy through the arrangement of charges. The capacitance value, normalized by the mass of the active material deposited on the single electrode, is called specific capacitance (C_{sp}) which is used to evaluate and compare the performance of the electrode materials. Generally, two electrochemical tests, such as cyclic voltammetry (CV) and galvanostatic charge–discharge (GCD), are used to calculate the C_{sp} values.

7.5.3 *Cyclic Voltammetry*

In CV, we can quantitatively and qualitatively measure the electrochemical performance of the electrode materials. In this technique, a fixed potential is applied to the working electrode with respect to the reference electrode. The potential window can be fixed based on the operating stability of the electrolytes. The current at the working electrode versus the applied voltage (i.e., the working electrode potential) is plotted to give the cyclic voltammogram trace for the calculation of capacitance. Ideally, the CV curves of the capacitor are rectangular in nature, however, EDLCs show deformed rectangular-shaped CV. Similarly, in the case of pseudocapacitive materials, oxidation and reduction peaks can be observed due to the faradaic reactions. From this, C_{sp} values can be evaluated. The specific capacitance can be calculated as given in the following equation:

$$C_{sp} = \frac{A}{FVm} \qquad (7.1)$$

where A is the area under the curve, F is the scan rate, V is the voltage window, and m is the mass of the active material in a single electrode.

At lower scan rates, the CV loops are rectangular in nature due to the proper use of available micropores in the active materials. However, with increasing scan rates, the ideal behavior is distorted. At higher scan rates, the ion transport is limited to the surface of the electrodes and does not utilize the pores of the active materials.

Whereas in case of pseudocapacitive materials, the charge storage mechanism is through faradaic reaction on the surface or near region of the electrode, which leads to the oxidation and reduction peaks in the CV and the charge stored depends on the potential.

7.5.4 Galvanostatic Charge–Discharge

GCD is the best and most proper method to measure the specific capacitance values. Here, a constant current is applied between the electrodes, and the potential is measured with respect to time. The capacitor is charged to a fixed potential, and monitoring the discharging by applying an opposite current is done to assess the capacitance. Similar to CV, EDLC, and pseudocapacitors have different shapes for the GCD curves. While EDLC materials show linear charge–discharge curves, pseudocapacitor has variations due to the redox reactions occurring at the electrodes. The specific capacitance values can be calculated from the slope of the discharging curve as given in the following equation:

$$C_{sp} = \frac{2i}{m\left(\Delta v \middle/ \Delta t\right)} \tag{7.2}$$

where i is the average cathodic current, $\Delta v/\Delta t$ is the slope of the discharge curve, and m is the mass of active material in one electrode.

Similar to CV, the specific capacitance decreases with increasing current density values in GCD. At higher current density, the charge–discharge is rapid and the proper usage of pores in the active materials is hindered [Ban *et al.*, 2013].

7.5.5 Electrochemical Impedance Spectroscopy

Electrochemical impedance spectroscopy (EIS) can be used to calculate the internal resistance, capacitance, and other electrical characteristics of the electrode materials. In this method, a small amplitude (5–10 mV) alternating current over the frequency range of 0.01 Hz to 100 kHz is applied to the supercapacitors. The response of the capacitor is plotted on the Nyquist plot (imaginary resistance (Z'') versus real resistance (Z')).

The shift in the real axis in the Nyquist plot means the x-intercept corresponds to the ESR, which gives the idea about the solution resistance of the electrolytes. The EDR corresponds to the charge-transfer resistance which depends on the electronic conductivity of the electrode materials. For an ideal capacitor, the vertical slope corresponds to the capacitance value which is invariant under frequency. In Fig 7.3, the portion of the supercapacitors is divided into two distinct regions.

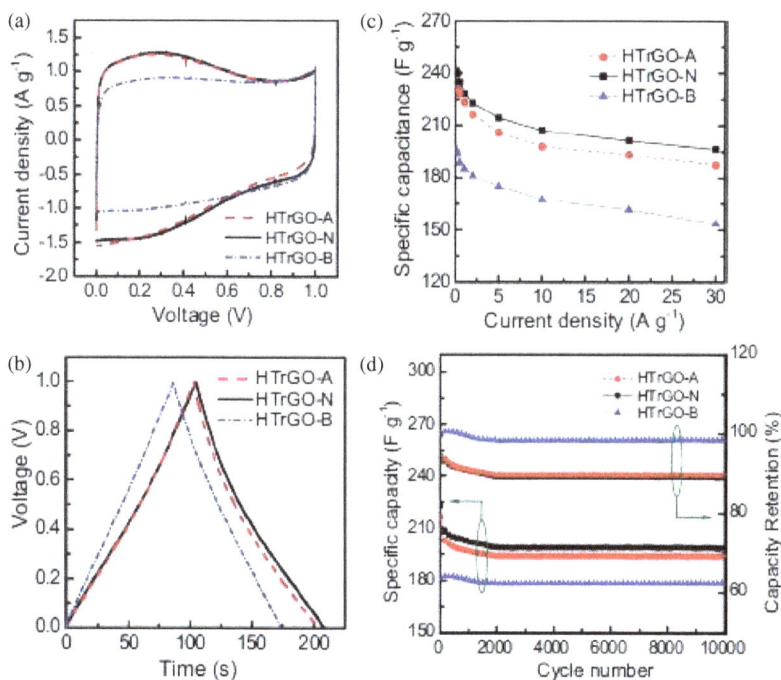

Fig 7.3 Schematic diagram of charging and discharging of the supercapacitor (a) cyclic voltammograms for hydrothermally rGOs (HTrGO) with different pH at a scan rate of 10 mV s⁻¹, (b) galvanostatic charge–discharge curve for HTrGOs at a constant current density of 1 A g⁻¹, (c) specific capacitance of HTrGOs at different current densities, and (d) cycling performance (10,000 charge–discharge cycles at a constant current density of 2 A g⁻¹). Reprinted with permission from Bai *et al.* [2013]. Copyright (2021) Elsevier.

The first one is the region with a 45° slope, encountered at a higher frequency. This portion of the graph corresponds to the diffusion process and is related to the diffusion of ions into the pores, which can be modeled using resistance and capacitance. In the lower frequency, the graph is almost a straight line showing the capacitive behavior [Kim *et al.*, 2015].

7.6 Graphene–MoS$_2$ Supercapacitor Applications

Zhu *et al.* [2011] fabricated porous graphene and 1-butyl-3-methyl-imidazolium tetrafluoroborate (BMIM BF4)/AN electrolyte-based supercapacitor and achieved a high specific capacitance of 166 F g^{-1} and a specific energy of 70 Wh kg^{-1}. It works in a wide potential range of 3.5 V. The high specific area of synthesized porous graphene was 2400 m^2 g^{-1}. In another work, a solid-state supercapacitor was fabricated by Liu *et al.* [2010] using mesoporous graphene. Celgard membrane and 1-ethyl-3-methyldizolium tetrafluoroborate (EMIMBF4) electrolytes were used to separate the sandwich layer of mesoporous graphene. The specific capacitance obtained for the supercapacitor device was 100–250 F g^{-1} at a potential window of 4 V and a current density of 1 A g^{-1}. Binders were used here for the fabrication of electrodes. However, binder-free flexible electrodes can reduce the cost of the device. Graphene was prepared using alkaline hydrothermal reduction of GO. The specific capacitance of 145 F g^{-1} was obtained in organic electrolytes and was around 95 F g^{-1} in ionic electrolytes. 1M lithium bis(trifluoromethane sulfonyl)imide (LiTFSI) in acetonitrile was used as organic electrolyte and ethylmethylimidazolium bis(trifluoromethane sulfonyl)imide (EMITFSI) as ionic electrolyte [Perera *et al.*, 2012]. Authors reported the highest specific capacitance value for hydrothermally reduced graphene-oxide-based supercapacitors. By hydrothermal reduction process, three kinds of rGOs were prepared and their pseudocapacitive performances were analyzed using full-cell supercapacitor devices. The specific capacity 185, 225, and 230 F g^{-1} were the values obtained for graphene produced in basic, acidic, and neutral solutions, respectively (at a current density of 1A g^{-1}) [Bai *et al.*, 2013]. Their high thermal stability, high theoretical basic capacitance, low cost, low toxicity, eco-friendly, and natural abundance are the major attractions of metal-oxide superca-pacitors. Supercapacitor electrodes with excellent electrochemical performance can be obtained by combining graphene with various metal oxides. Some of the basic electrodes for supercapacitor applica-tions are NiO, FeO, CoO, RuO, etc.

Rakhi *et al.* [2011] reported SnO_2/GNs-, MnO_2/GNs- and RuO_2/GNs-based supercapacitor electrodes (Fig 7.4). The specific capacitance value obtained for SnO_2/GNs composites was 195 F g^{-1} and that of RuO_2/GNs was 365 F g^{-1} at 20 mV s^{-1} under two-electrode configuration. The supercapacitors exhibited higher energy and power densities for all three composites. The combination of graphene with metal oxides enhances electrical conductivity and enables the transportation of ions and molecules while storing the charge. The variable oxidation state of the metal is the factor that determines the suitability of a metal oxide for effective charge storage. In addition to single metal oxides such as NiO, MnO, RuO, and CoO, binary metal oxides ($NiCo_2O_4$, $NiMn_2O_4$) and their combina-tions with graphene or rGO can also be utilized for charge storage [Chen *et al.*, 2010] (Fig 7.5).

Graphene can serve as an efficient supercapacitor electrode when combined with conductive polymers (CPs). Some of the CPs are polypyrrole (PPy), polyaniline (PANI), and poly(3,4-ethylenedioxy thiophene) polystyrene sulfonate (PEPOT:PSS) [Dong *et al.*, 2012]. They offer pseudocapacitance, owing to the very fast redox reaction with electrolyte. Cong *et al.* [2013] fabricated a supercapacitor using flexible graphene/PANI. The specific capacitance of the device is 763 F g^{-1} upon electropolymerizing PANI nanorods on the above graphene paper. This is a cost-effective method and offers better cycling stability (82% after 1000 cycles). Chen *et al.* [2013] reported a 36–92% increase in capacitance and a very efficient cycling perfor-mance (>50,000 cycles) for a redox electrolyte in combination with PANI/graphene supercapacitor. The only significant demerit of CP is its short-term stability, which can be resolved by combining it with graphene and thereby attaining long-term cyclic stability which is desirable and compulsory for supercapacitor devices. Graphene

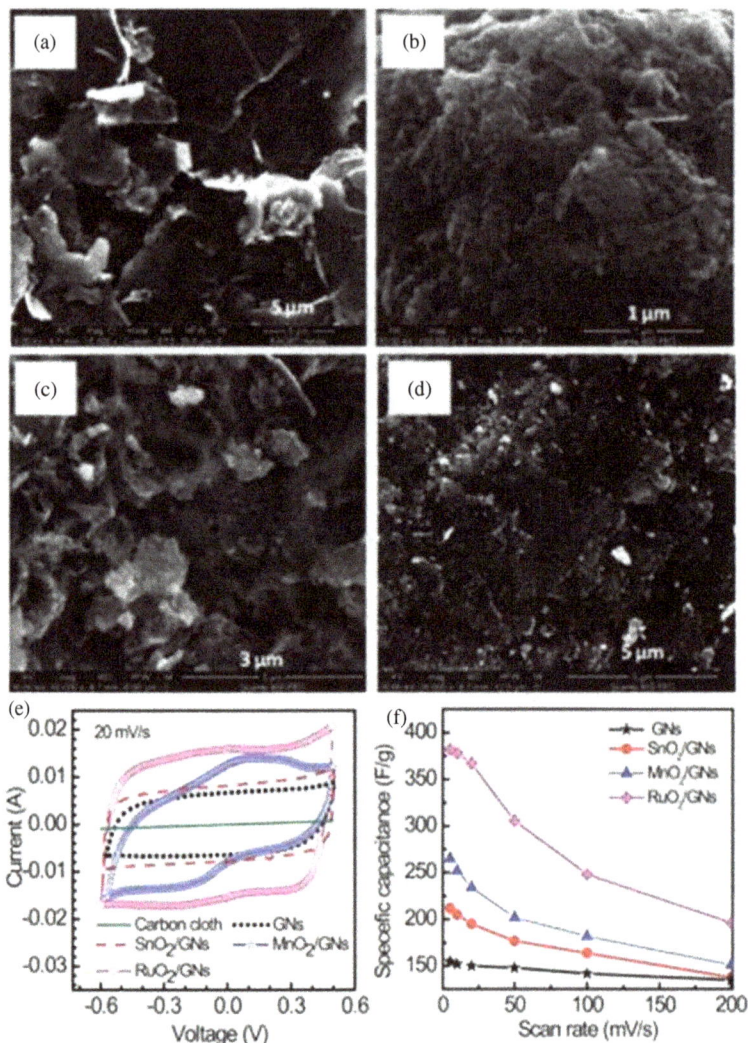

Fig 7.4   Schematic of the charging and discharging of the supercapacitor SEM images of (a) GNs, (b) SnO_2/GNs, (c) MnO_2/GNs, and (d) RuO_2/GNs. Electrochemical performance of (e) comparison CV curves of the metal-oxide dispersed GNs at 20 mV s^{-1} and (f) variation of specific capacitance with scan rate. Reprinted with permission from Rakhi *et al.* [2011]. Copyright (2021) Royal Society of Chemistry.

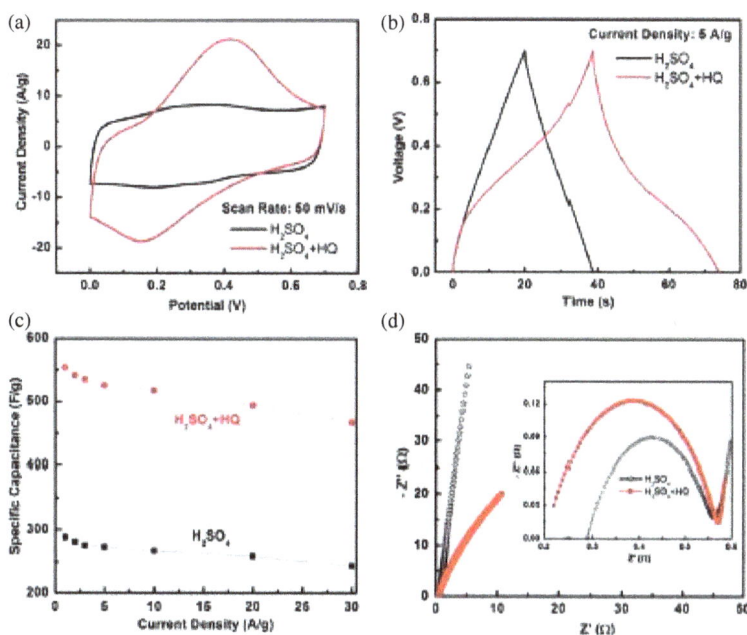

Fig 7.5 Schematic of the charging and discharging of the supercapacitor (a) CV, (b) GCD, (c) specific capacitance, and (d) electrochemical impedance of H_2SO_4 and H_2SO_4-hydroquinone (HQ) electrolytes for rGO-PANI supercapacitors. Reprinted with permission from Chen *et al.* [2013]. Copyright (2021) Royal Society of Chemistry.

improves total conductivity by working as a conductive medium for the transfer of charge, whereas metal oxides/hydroxides or CPs provide pseudocapacitance [Karthikeyan *et al.*, 2012].

Numerous studies have been conducted to study the combination of graphene with metal oxides and polymer nanocomposites. Currently, many researchers are focused on the supercapacitor application of the graphene–MoS₂ combination, owing to the fascinating properties of this combination. Herein, we discuss the recent advances in the field of supercapacitors based on graphene–MoS₂ combination.

Among the transition metal dichalcogenides (TMDs), MoS₂ has been widely studied due to its sheet-like structure, chemical and thermal stability, flexibility, excellent conductivity, high surface-to-volume ratio, and high theoretical capacitance of 1500 F g^{-1}. Among the pseudocapacitive materials, MoS$_2$ shows one of the highest theoretical capacitances. MoS$_2$ as an electrode material shows some reversible electrochemical reaction that contributes to the capacitance.

$$MoS_2 + K^+ + e^- \leftrightarrow MoS + SK^+ \qquad (7.3)$$

In recent years, the researchers have been mainly focusing on the tuning of MoS$_2$ into different morphologies, and developing different phases for different applications. For supercapacitor applications, we need to develop high surface area materials to enhance specific capacitance without compromising electronic conductivity. There are several synthetic strategies reported for MoS$_2$, such as exfoliation of a few layers from bulk using solvents, mechanical forces, chemical vapor deposition, pulsed laser deposition, microwave irradiation techniques, and hydrothermal synthesis [Ramadoss *et al.*, 2014]. Hydrothermal synthesis is the main technique upon which researchers depend due to its easiness to synthesize different morphologies, such as nanowires, nanospheres, nanoflowers, nanorods, and nanofibers in bulk amounts, by changing temperature, reaction time, solvents, and pH of the solution. The flower-shaped structure of MoS$_2$ is commonly suggested for storage applications because it shows a high surface area for the storage of the electrolyte ions [Liu *et al.*, 2019]. Mesoporous hexagonal phase MoS$_2$ can be synthesized using single- and multiple-step hydrothermal techniques [Huang *et al.*, 2014]. It showed a specific capacitance of 403 F g^{-1} at 1 mV s^{-1} and 80% cycling stability over 2000 galvanostatic charge–discharge in a symmetric configuration. They have shown the capacitance change with electrolytes, but there is not much difference in the C_{sp}, because they have studied both in a water-based electrolyte. Several studies on supercapacitors were based on hydrothermally synthesized MoS$_2$ and have shown promising results in both three-electrode configuration and real symmetric capacitor configuration [Balasingam *et al.*, 2017]. However, there are some problems that need to be addressed in the area of capacitor research. It includes the inferior stability of the electrode due to the aggregation and reduction of the active material during the charge–discharge cycles. The stability problem can be controlled by modifying the structure and morphology of the active material. To overcome this challenge, direct deposition of MoS_2 over current collectors can be adopted, which will give better stability to the electrodes. A high rate and stable supercapacitor can be developed through in situ preparation of electrode material over the conducting. Instead of making slurry and depositing the material, using drop-casting has some disadvantages. The main problem in drop-casting method is the resistance between the supporting electrode and electrode materials, and also there is a chance of the material getting collapsed when electrolyte ions are added. To overcome these, people commonly use nonconducting polymers as a binder that can hinder the capacitance property of the active materials. The in situ preparation of material over the electrode helps in overcoming the problems. This can be done easily and cost-effectively in hydrothermal synthesis compared to other techniques. The in situ preparation of material in hydrothermal synthesis also helps in having high mass-load electrodes which will eliminate the negative contribution of current collectors [Manuraj *et al.*, 2020]. To obtain the best performance from electrode material, it is important that it should be optimized over different electrolytes (acidic, basic, and neutral). MoS_2 is known to show better capacitance and stability in neutral electrolytes [Kanaujiya *et al.*, 2019].

The hexagonal phase of MoS_2 is widely reported for its storage capacity and for other semiconducting applications due to its easiness of synthesis and also because the hexagonal phase is the most stable phase of MoS_2. However, for storage applications, intrinsic conductivity is also an important parameter. Most of the studies conducted with MoS_2-based supercapacitors are with hexagonal structures (semiconducting) or with hybrid phases [Wang *et al.*, 2017]. Tetragonal (metallic) MoS_2 still needs to be considered for synthesis and storage application due to its high intrinsic conductivity. Because of the poor conductivity of the MoS_2, graphene or other conducting material were used to improve conductivity. The authors developed metallic, multilayer MoS_2 with H_2O as a separator. The separation between each layer is maintained by the nanochannels present between the

H_2O molecules and helps in the easy passage of electrolyte ions inside the layers, and in the agglomeration and enlarging of the surface area. Supercapacitor performance in different electrolytes was also studied. Here the high conductivity of the MoS_2 helps to improve the working of the capacitor even in high scan rates up to 10 V s^{-1} with promising specific capacitance. The specific capacitance of the M-MoS_2-H_2O system with symmetric configuration shows a scan rate of 50 mV s^{-1} at 249 F g^{-1}. The authors showed the dependence of nanochannels on the ion accessibility of the different electrolytes [Geng *et al.*, 2017] (Fig 7.6).

The synthesis of large-scale materials can be easily done by hydrothermal method but when it comes to device fabrication, there are some challenges such as uniformity of the electrode, control over the growth of the layer, and quality of the film. In order to overcome these problems, we have to choose another film fabrication technique like magneton sputtering [Choudhary *et al.*, 2015]. The direct deposition technique helps increase the conductivity and also overcomes the interface resistance between the supporting current collector and electrode materials. The conductivity can be reduced by the commonly used supporting polymer binders. They also hinder the movement of the electrolyte ions between the materials. This method helps them to improve the capacitance value that was already reported using the sputtering technique. Recently, the atomic layer deposition (ALD) technique was widely used for thin film technology. Using this techniques people developed electrode materials for energy storage applications. ALD is a self-limiting, two-stage thin film technique that helps make uniform thin films on 2D as well as 3D scaffolds [Nandi *et al.*, 2017].

Unlike hydrothermal or other techniques, exfoliation of MoS_2 can develop few-layers or single-layer 1T phase. The major advantage of this technique is that we can improve the in-plane conductivity as well as the mobility of the material and also the distance between the layers of the material, thereby enhancing the intercalation of the ions between the layers. It helps the supercapacitor performance with

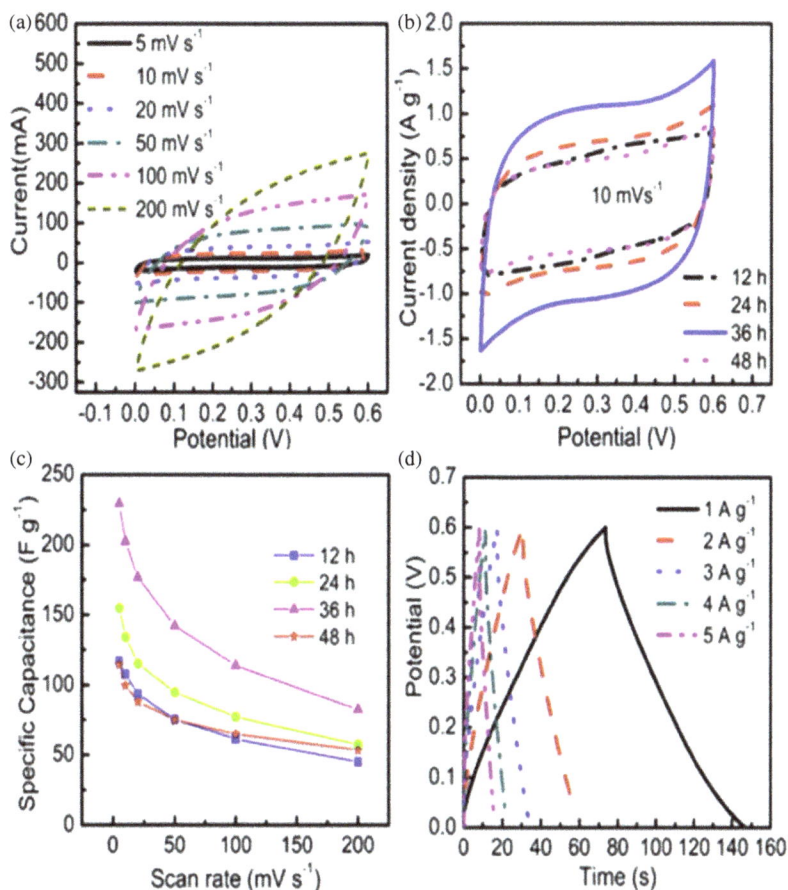

Fig 7.6 (a) The CV curves of symmetric supercapacitors based on MoS₂ synthesized directly over nickel foam at 36 h, (b) comparison of CV curves of symmetric supercapacitors at a scan rate of 10 mV s⁻¹, (c) variation of specific capacitance as a function of scan rate, (d) the galvanostatic charge–discharge of symmetric supercapacitors based on MoS₂ synthesized directly over nickel foam at 36 h, (e) comparison of charge–discharge curves of symmetric supercapacitors at 1 A g⁻¹, and (f) variation of specific capacitance as a function of current density. Reprinted with permission from Manuraj *et al.*, [2020]. Copyright (2021) Elsevier.

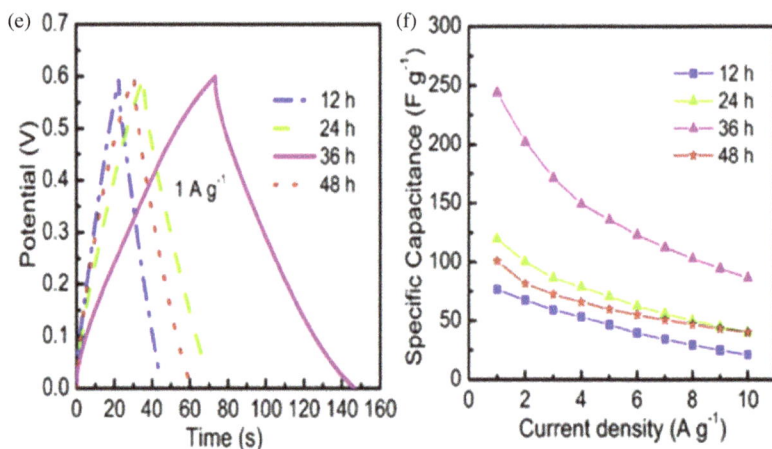

Fig 7.6 (*Continued*)

high-rate capability and delivers high capacitance. There are different exfoliation techniques such as ball-milling, scotch tape, and liquid exfoliation. Intercalation of Li ions has got much attention and also intercalation pseudocapacitance shows high capacitance compared to conventional supercapacitors with high-rate capability. This is because the ions that intercalate into the layers or active tunnels of the active material will be accompanied by a faradaic charge-transfer reaction without any phase change. But the problem in these types of synthesis of electrode material is the agglomeration leading to a lesser mass of materials through separation, and also the energy density values are less compared to commercial values [Burke, 2007]. We can achieve the energy density compared to commercial one by two approaches, either by fabricating the electrodes without using any polymer binders or changing the electrolytes to ionic or organic electrolytes. In the case of exfoliation technique, it is difficult to synthesize binder-free electrode materials, so we have to concentrate on the use of electrolytes [Pandey *et al.*, 2015].

Electrochemical exfoliation is a new technique that solves most of the problems that we face with other exfoliation techniques. Cathodic and anodic exfoliations are the two major strategies used for this depending on the intercalation of cations or anions. Cathodic exfoliation involves Li-based salts with organic solvents, whereas in anodic exfoliation, sulfate-based aqueous electrolytes are used as the medium for the intercalation. There are some drawbacks to the techniques of separating the materials from bulk. In the case of cathodic exfoliation, the Li ions intercalate to change the phase of the active materials, whereas, in the case of anodic exfoliation, oxygen in the water molecules attacks the active material and therefore may change the material into other oxide or hydroxide forms. In order to overcome these problems, Sergio Garcia and his coworkers introduced heavy alkali metal-based salts in place of Li. These water-soluble salts may reduce the use of organic solvents thereby making this strategy cost-effective. The alkali metals are larger in size compared to Li, thus reducing the number of ions intercalated into the layer and limiting the number of electrons injected. This helps block the phase transformation of MoS_2 [Garcia-Dali *et al.*, 2019].

Graphene is a well-known EDLC-based electrode material. Due to its excellent stability and conductivity, it has been used as an electrode material for supercapacitors. However, the aggregation of graphene nanosheets makes it using the surface area difficult as expected from the theoretical data and also it is difficult to disperse or dissolve even after continuous ultrasonication. The discrete nature of exfoliated graphene sheets stacked loosely in the porous current collectors may lead to high contact resistance and unfavorably decrease the performance of supercapacitors [Xu *et al.*, 2011]. Therefore, surface modification of graphene improves its performance as an electrode material. To improve the capacitance, MoS_2-like graphene analog materials can be used which can provide two types of charge storage mechanisms in supercapacitors, which will increase the capacitance and energy density. The stability and conductivity are the main issues in MoS_2-based supercapacitors but using graphene with MoS_2 may overcome these deficiencies. The physical mixture of few layers of MoS_2 from exfoliation with graphene can be used to prepare thin membrane electrode materials which show low resistance over electrolyte ions due to the intercalation and de-intercalation of ions between few layers of the materials during electrochemical process which may also improve the stability of the capacitor. The mixture of graphene and MoS_2 membrane shows a capacitance of 11 mF cm^{-2} at 5 mV s^{-1} (Fig 7.7).

The composite shows improvement in capacitance after certain charge–discharge cycles which is due to proper ion intercalation [Bissett *et al.*, 2015]. Microwave-assisted and hydrothermal synthesis of MoS_2-graphene composites give better performance due to their overlapping or coalescing structure. Covalently bonded MoS_2 over graphene oxide gives better stability. [Xu *et al.*, 2018]. Three-dimensional nanospheres of MoS_2–graphene composite prepared through hydrothermal synthesis show improved capacitance compared with graphene (35 F g^{-1} at 1 A g^{-1}) and MoS_2 (120 F g^{-1} at 1 A g^{-1}) around 243 F g^{-1} at 1 A g^{-1}. The result shows almost double

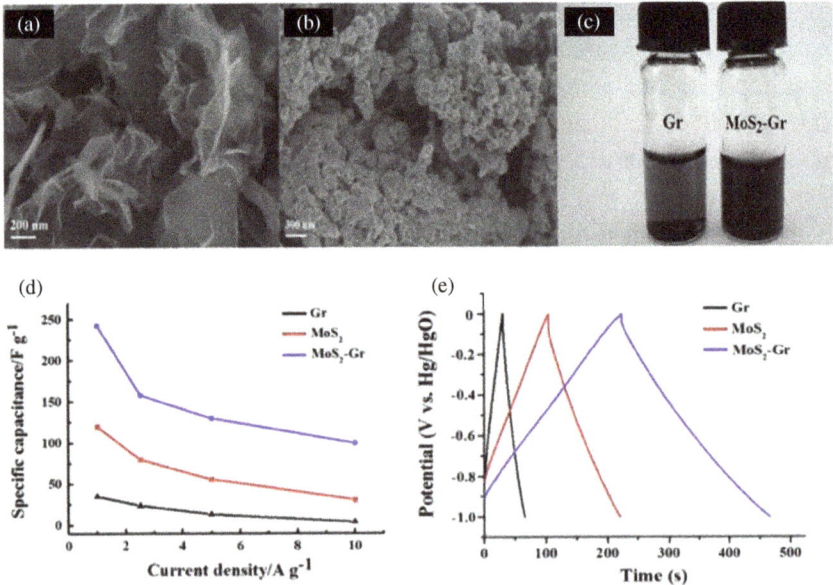

Fig 7.7   Schematic of the charging and discharging of the supercapcitors (a) SEM images of Gr, (b) MoS_2-Gr, (c) the photograph of Gr and MoS_2-Gr dispersed solution after left to stand for two days, (d) specific capacitance of the MoS_2-Gr composites at different current densities of 1, 2.5, 5, and 10 A g^{-1} in 1 M Na_2SO_4, and (e) GCD curves of MoS_2, Gr and MoS_2-Gr composites at 1 A g^{-1}. Reprinted with permission from Huang *et al.* [2013]. Copyright (2021) Elsevier.

the value of capacitance and higher energy and power density of 73.5 Wh kg^{-1} at a power density of 19.8 kW kg^{-1} [Huang *et al.*, 2013]. The development of two different nanostructured materials through hydrothermal techniques helps to improve proper charge transfer and decrease the charge-transfer resistance. MoS$_2$–graphene binder-free electrodes show a high capacitance of around 387.6 F g^{-1} at 1.2 A g^{-1}. Also, the incorporation of graphene improves stability without any loss over 1000 cycles [Saraf *et al.*, 2018].

7.7 Conclusion

Electrochemical supercapacitors are emerging energy storage devices with many exciting attributes such as high power density and long cycle life. They are widely employed in hybrid vehicles and high-power applications, but their energy density is considerably lower than that of rechargeable batteries. The researchers are putting diligent efforts to find new materials with high capacitance and a wide potential window. The electrode material should have a high surface area, pore size distribution, and pore length for the diffusion of ions. Nanomaterials that can satisfy all these properties are to be carefully selected to overcome the existing limitations. Graphene was found to be suitable for electrode materials due to its high surface area and good electronic conductivity. But here also, low energy density remains a defect. Developing composite materials with highly stable graphene is a relevant idea to resolve this problem to a greater extent. MoS$_2$, also known as graphene analogue (due to its sheet-like structure), can be employed for this purpose. The pseudocapacitive property of these materials will provide improved specific capacitance, nearly solving the problem of low energy density. Different morphologies of these materials can be developed using different synthesis strategies. The graphene–MoS$_2$ combination provides supercapacitors of exceptional stability.

References

Bai, Y., Rakhi, R. B., Chen, W. & Alshareef, H. N. (2013). Effect of pH-induced chemical modification of hydrothermally reduced graphene oxide on supercapacitor performance. *J. Power Sources* 233, pp. 313–319.

Balasingam, S., Lee, M., Kim, B., Lee, J. & Jun, Y. (2017). Freeze-dried MoS$_2$ sponge electrodes for enhanced electrochemical energy storage. *Dalton Trans.* 46(7), pp. 2122–2128.

Ban, S., Zhang, J., Zhang, L., Tsay, K., Song, D. & Zou, X. (2013). Charging and discharging electrochemical supercapacitors in the presence of both parallel leakage process and electrochemical decomposition of solvent. *Electrochim. Acta* 90, pp. 542–549.

Bissett, M. A., Kinloch, I. A. & Dryfe, R. A. W. (2015). Characterization of MoS$_2$-graphene composites for high-performance coin cell supercapacitors. *Acs Appl. Mater. Inter.* 7(31), pp. 17388–17398.

Burke, A. (2007). R&D considerations for the performance and application of electrochemical capacitors. *Electrochim. Acta* 53(3), pp. 1083–1091.

Chen, W., Rakhi, R. B. & Alshareef, H. N. (2013). Capacitance enhancement of polyaniline coated curved-graphene supercapacitors in a redox-active electrolyte. *Nanoscale* 5(10), pp. 4134–4138.

Chen, S., Zhu, J., Wu, X., Han, Q. & Wang, X. (2010). Graphene oxide-MnO$_2$ nanocomposites for supercapacitors. *ACS Nano* 4(5), pp. 2822–2830.

Choudhary, N., Patel, M., Ho, Y. H., Dahotre, N. B., Lee, W., Hwang, J. Y. & Choi, W. (2015). Directly deposited MoS$_2$ thin film electrodes for high performance supercapacitors. *J. Mater. Chem. A* 3(47), pp. 24049–24054.

Cong, H.-P., Ren, X.-C., Wang, P. & Yu, S.-H. (2013). Flexible graphene–polyaniline composite paper for high-performance supercapacitor. *Energy Environ. Sci* 6(4), pp. 1185–1191.

Conway, B. E. (1991). Transition from "Supercapacitor" to "Battery" behavior in electrochemical energy storage. *J. Electrochem. Soc.* 138(6), pp. 1539–1548.

Dong, X. C., Xu, H., Wang, X. W., Huang, Y. X., Chan-Park, M. B., Zhang, H., Wang, L. H., Huang, W. & Chen, P. (2012). 3D graphene-cobalt oxide electrode for high-performance supercapacitor and enzymeless glucose detection. *Acs Nano* 6(4), pp. 3206–3213.

Garcia-Dali, S., Paredes, J. I., Munuera, J. M., Villar-Rodil, S., Adawy, A., Martinez-Alonso, A. & Tascon, J. M. D. (2019). Aqueous cathodic exfoliation strategy toward solution-processable and phase-preserved MoS_2 nanosheets for energy storage and catalytic applications. *ACS Appl. Mater. Interfaces* 11(40), pp. 36991–37003.

Geng, X., Zhang, Y., Han, Y., Li, J., Yang, L., Benamara, M. & Zhu, H. (2017). 2D water-coupled metallic MoS_2 with nanochannels for ultrafast supercapacitor. *Nano Lett.* 17, pp. 1825–1832.

Huang, K. J., Wang, L., Liu, Y. J., Liu, Y. M., Wang, H. B., Gan, T. & Wang, L. L. (2013). Layered MoS_2-graphene composites for supercapacitor applications with enhanced capacitive performance. *Int. J. Hydrogen Energ.* 38(32), pp. 14027–14034.

Huang, K.-J., Zhang, J.-Z., Shi, G.-W. & Liu, Y.-M. (2014). Hydrothermal synthesis of molybdenum disulfide nanosheets as supercapacitors electrode material. *Electrochim. Acta* 132, pp. 397–403.

Jiang, Y. & Liu, J. (2019). Definitions of pseudocapacitive materials: A brief review. *Energy Environ. Mater.* 2(1), pp. 30–37.

Kanaujiya, N., Kumar, N., Sharma, Y. & Varma G. D. (2019). Probing the electrochemical properties of flower like mesoporous MoS_2 in different aqueous electrolytes. *J. Electron. Mater.* 48(2), pp. 904–915.

Karthikeyan, K., Kalpana, D., Amaresh, S. & Lee, Y. S. (2012). Microwave synthesis of graphene/magnetite composite electrode material for symmetric supercapacitor with superior rate performance. *RSC Adv.* 2(32), pp. 12322–12328.

Kim, B. K., Sy, S., Yu, A. & Zhang, J. (2015). Electrochemical supercapacitors for energy storage and conversion. in: Handbook of clean energy systems. pp. 1–25.

Liu, X., Liu, L., Wu, Y., Wang, Y., Yang, J. & Wang, Z. (2019). Rosette-like MoS_2 nanoflowers as highly active and stable electrodes for hydrogen evolution reactions and supercapacitors. *RSC Adv.* 9(24), pp. 13820–13828.

Liu, C., Yu, Z., Neff, D., Zhamu, A. & Jang, B. Z. (2010). Graphene-Based Supercapacitor with an ultrahigh energy density. *Nano Lett.* 10(12), pp. 4863–4868.

Manuraj, M., Kavya Nair, K. V., Unni, K. N. N. & Rakhi, R. B. (2020). High performance supercapacitors based on MoS_2 nanostructures with near commercial mass loading. *J. Alloys Compd.* 819, p. 152963.

Nandi, D. K., Sahoo, S., Sinha, S., Yeo, S., Kim, H., Bulakhe, R. N., Heo, J., Shim, J. J. & Kim, S. H. (2017). Highly uniform atomic layer-deposited MoS_2@3D-Ni-Foam: A novel approach to prepare an electrode for supercapacitors. *ACS Appl. Mater. Interfaces* 9(46), pp. 40252–40264.

Pandey, K., Yadav, P. & Mukhopadhyay, I. (2015). Electrochemical and electronic properties of flower-like MoS_2 nanostructures in aqueous and ionic liquid media. *RSC Adv.* 5(71), pp. 57943–57949.

Perera, S. D., Mariano, R. G., Nijem, N., Chabal, Y., Ferraris, J. P. & Balkus, K. J. (2012). Alkaline deoxygenated graphene oxide for supercapacitor applications: An effective green alternative for chemically reduced graphene. *J. Power Sources* 215, pp. 1–10.

Rakhi, R. B., Chen, W., Cha, D. & Alshareef, H. N. (2011). High performance supercapacitors using metal oxide anchored graphene nanosheet electrodes. *J. Mater. Chem.* 21(40), pp. 16197–16204.

Ramadoss, A., Kim, T., Kim, G.-S. & Kim, S. J. (2014). Enhanced activity of a hydrothermally synthesized mesoporous MoS_2 nanostructure for high performance supercapacitor applications. *New J. Chem.* 38(6), pp. 2379–2385.

Ren, X., Fan, H., Ma, J., Wang, C., Zhang, M. & Zhao, N. (2018). Hierarchical Co_3O_4/PANI hollow nanocages: Synthesis and application for electrode materials of supercapacitors. *Appl. Surf. Sci.* 441, pp. 194–203.

Saraf, M., Natarajan, K. & Mobin, S. M. (2018). Emerging robust heterostructure of MoS_2-rGO for high-performance supercapacitors. *ACS Appl. Mater. Interfaces* 10(19), pp. 16588–16595.

Tiwari, P., Janas, D. & Chandra, R. (2021). Self-standing MoS_2/CNT and MnO2/CNT one dimensional core shell heterostructures for asymmetric supercapacitor application. *Carbon* 177, pp. 291–303.

Wang, D., Xiao, Y., Luo, X., Wu, Z., Wang, Y.-J. & Fang, B. (2017). Swollen Ammoniated MoS_2 with 1T/2H Hybrid phases for high-rate electrochemical energy storage. *ACS Sustainable Chem. Eng.* 5, pp. 2509–2515.

Winter, M. & Brodd, R. J. (2004). What are batteries, fuel cells, and supercapacitors? *Chem. Rev.* 104(10), pp. 4245–4270.

Xu, X., Wu, L., Sun, Y., Wang, T., Chen, X., Wang, Y., Zhong, W. & Du, Y. (2018). High-rate, flexible all-solid-state super-capacitor based on porous aerogel hybrids of MoS_2/reduced graphene oxide. *J. Electroanal. Chem.* 811, pp. 96–104.

Xu, B., Yue, S., Sui, Z., Zhang, X., Hou, S., Cao, G. & Yang, Y. (2011). What is the choice for supercapacitors: Graphene or graphene oxide? *Energy Environ. Sci.* 4(8), pp. 2826–2830.

Zhu, Y., Murali, S., Stoller, M. D., Ganesh, K. J., Cai, W., Ferreira, P. J., Pirkle, A., Wallace, R. M., Cychosz, K. A., Thommes, M., Su, D., Stach, E. A. & Ruoff, R. S. (2011). Carbon-based supercapacitors produced by activation of graphene. *Science* 332(6037), p. 1537.

https://doi.org/10.1142/9789811283406_0008

Chapter 8

Carbon-Based and TMDs-Based Air Cathode for Metal–Air Batteries

Prince Kumar Maurya and Ashish Kumar Mishra

School of Materials Science and Technology
Indian Institute of Technology (Banaras Hindu University),
Varanasi 221005
Email: akmishra.mst@iitbhu.ac.in

Abstract

Metal–air batteries (MABs) have emerged as a promising solution for high-energy density storage in applications such as electric vehicles and smart grids. However, several challenges hinder the successful implementation of MABs in the market. This chapter focuses on addressing these challenges and enhancing the performance of MABs through innovative material selection, design, and manufacturing techniques. To overcome this, researchers are investigating nonmetallic (heteroatom)-doped carbon materials and TMDs-based nanocomposites as electrocatalysts for air cathode in MABs.

8.1 Introduction

The anticipated rise in electricity demands the exploration of alternative energy sources such as solar, hydropower, and wind energy. To meet this demand, it is imperative to advance the development of high-energy storage systems. These systems will allow for excess energy generated by renewable sources to be stored and utilized during their high demands by providing a consistent power source [Cheng & Chen, 2012; Wang *et al.*, 2021]. Lead-acid batteries were popular for electric vehicle propulsion in the 1990s, but their low energy density limited their usage. Currently, most electric vehicles, especially cars, are adopting lithium (Li)-ion batteries for propulsion. These batteries have a limited range of 160 km per charge, making up a significant portion (65%) of their cost due to lower energy density ranging from 100 to 200 Wh kg^{-1} compared with gasoline with energy density of 13,000 Wh kg^{-1} [Dunn *et al.*, 2011; Rao & Wang, 2011; Taniguchi *et al.*, 2001]. Pb-acid, Ni-Cd, and Li-ion batteries are prevalent in portable electronics and electric vehicles. Although there have been some positive advancements, particularly with Li-ion batteries, they remain insufficient for enhancing the range of electric vehicle propulsion due to their high cost, environmental hazards, and practical energy density differing from theoretical expectations.

To perform long run, electric vehicles should have great specific energy and power density. The attainment of a long-driving range is positively correlated with a high specific-energy density, whereas the ability for rapid acceleration and hill-climbing is proportional to the presence of a high specific-power density [Chau *et al.*, 1999]. Due to their open battery design, utilizing air as a reagent, metal–air batteries (MABs) have significantly higher specific capacities. However, these systems face significant challenges that must be resolved before they can be utilized in practical applications. MABs with open-cell structure allow ambient oxygen to flow toward cathode material, which differ from closed/packed Li-ion batteries. They are composed of the air cathode, metal anode, electrolyte, and separator. Various electrocatalysts such as porous carbon materials have been extensively investigated as air electrodes and metals such as Na, Fe, Li, and Zn as anode [Chang *et al.*, 2015]. A separator functions as an insulator that permits only ion transportation. Reduction reaction at cathode and oxidation reaction at anode take place during the entire discharge process. The anode metal dissolves in the electrolyte, while the air cathode induces an oxygen-reduction-reaction (ORR) mechanism and oxygen-evolution-reaction (OER) mechanism during the charging of the metal–air cell [Zhang *et al.*, 2016]. In the following sections, we will discuss multiple MABs based on carbon and TMD-based cathodes.

8.2 Working Principle of MABs

An electrochemical metal–air cell utilizes a metal anode and a cathode composed of porous nanostructures for ambient air. It can employ either an aqueous or aprotic electrolyte. During discharge, reduction occurs in the surrounding air at the cathode, while the metal anode undergoes oxidation. Metal–air electrochemical cells have a higher specific capacity and energy density than Li-ion batteries, making them a promising alternative for application in electric vehicles. However, the development and implementation of MABs face difficulties such as metal anodes, catalysts, and electrolytes. Despite their potential to replace Li-ion batteries, these issues have impeded their progress [Zhang *et al.*, 2016]. Fig 8.1(a) illustrates the schematic of an MAB. In this process, metal ions (M^+) are generated at the anode and flow through the electrolyte toward the cathode, where they react with O_2 to produce metal oxides (MO_{2x}). The following reactions (8.1–8.3) exhibit the process:

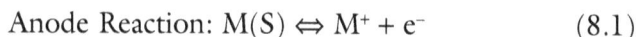

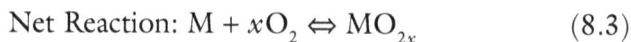

$$\text{Anode Reaction: } M(S) \Leftrightarrow M^+ + e^- \tag{8.1}$$

$$\text{Cathode Reaction: } M^+ + xO_2 + e^- \Leftrightarrow MO_{2x} \tag{8.2}$$

$$\text{Net Reaction: } M + xO_2 \Leftrightarrow MO_{2x} \tag{8.3}$$

The most widely used MABs are magnesium–air, lithium–air, potassium–air, sodium–air, and zinc–air batteries. The Li–air batteries boost the substantial energy density to ~5,200 Wh kg^{-1}, far exceeding the prevalent Li-ion batteries with a charge-to-mass ratio of 100–200 Wh kg^{-1}, making them a good choice for electric vehicles and electronic devices [Ma *et al.*, 2018]. The theoretical energy densities of various MABs such as magnesium–air, lithium–air, sodium–air,

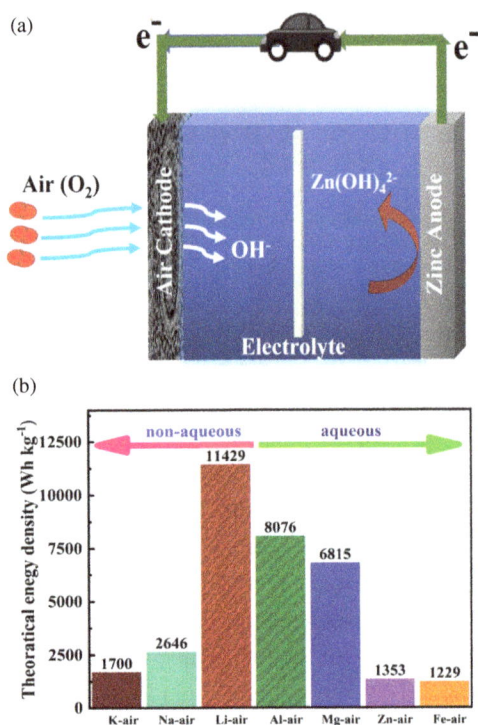

Fig 8.1   (a) Schematic diagram of a typical MAB and (b) theoretically estimated energy densities of various types of MABs. Adapted from Li and Lu [2017].

potassium–air, and zinc–air batteries are illustrated in Fig 8.1(b). Notably, these batteries possess a theoretically estimated energy density that is significantly greater than the conventional Li-ion batteries by a factor of 3–30 [Li & Lu, 2017].

8.2.1 *Electrochemical Study*

The electrochemical activity of electrocatalysts in terms of OER and ORR is vital for the performance, efficiency, and stability of MABs. The linear sweep voltammetry (LSV) measurements are commonly used to determine the Tafel slope, overpotential (η_{10}), number of

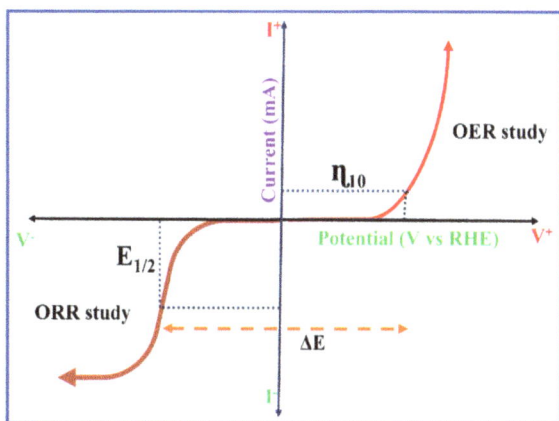

Fig 8.2   Schematic representation of LSV measurement for OER and ORR.

electrons (n) participating in oxygen-reduction/evolution kinetics, and the half-wave potential ($E_{1/2}$) of electrocatalysts for OER and ORR, as shown in Fig 8.2.

8.2.1.1 *ORR Analysis*

The ORR measurement provides insights into discharge reaction kinetics, mechanisms, and intermediates, aiding catalyst design and optimization. It also helps in assessing catalyst durability under harsh operating conditions, ensuring long-term device performance. LSV measurements are commonly used to determine the Tafel slope, number of electrons (n) participating in oxygen-reduction kinetics and the half-wave potential ($E_{1/2}$) of electrocatalysts for ORR [Huang *et al.*, 2015]. The Tafel slope is a parameter used to describe the electrochemical kinetics of a reaction and can be expressed as follows:

$$\eta = a + b\log J \qquad (8.4)$$

where η represents the overpotential (V or mV), a is intercepts, and b is the Tafel slope (V dec^{-1} or mV dec^{-1}). These parameters help in evaluating and comparing the performance of different materials or catalysts for the ORR. A lower Tafel slope and a higher $E_{1/2}$ potential in ORR indicate a more efficient electrocatalyst with a faster reaction rate.

To calculate the electron transfer number (n) and to gain a measurable quantity of the ORR process utilizing rotating disc electrode (RDE) data during the ORR, the Koutecky–Levich (K-L) Equations 8.5–8.7 are used [Shinagawa *et al.*, 2015].

$$J_D = \frac{1}{J_L} + \frac{1}{J_K} = \frac{1}{(B\omega^{1/2})} + \frac{1}{J_K} \tag{8.5}$$

$$B = 0.2nFC_oD_o^{\frac{2}{3}}\vartheta^{-\frac{1}{6}} \tag{8.6}$$

$$J_K = nFkC_o \tag{8.7}$$

where J_D and J_K are the measured disc current density and the kinetic-limiting current density, respectively, and ω is the angular velocity in revolutions per minute (rpm). D_o is the O_2 diffusion coefficient in 0.1 M KOH ($D_o = 1.9 \times 10^{-5}$ m² s⁻¹) and C_o represents the bulk O_2 concentration in 0.1 M KOH ($C_o = 1.26 \times 10^{-3}$ mol L⁻¹). The kinematic viscosity of the electrolyte is denoted as υ ($= 0.01$ cm² s⁻¹) and k is the ORR electron-transfer rate constant. Here, n stands for the number of transferred electrons and F denotes the Faradaic constant ($= 96485$ C mol⁻¹). When specifying the electrode rotation speed in rpm, the constant 0.2 is used and B can be calculated from the slope of the K-L equation. The calculated electron-transfer number (n) for the material indicates that the ORR follows a pseudo 4e⁻ or 2e⁻ reaction pathway [Ge *et al.*, 2015; Wu *et al.*, 2009]. The catalyst's ORR activity is further confirmed by scanning the disk electrode at the cathode level in the rotating ring-disk electrode experiment. The percentage of H_2O_2 and n during ORR can be computed using Equations (8.8, 8.9) [Antoine & Durand, 2000; Sharifi *et al.*, 2012].

$$\%H_2O_2 = 2J_R + (NJ_D + J_R) \tag{8.8}$$

$$n = \frac{4J_D}{(J_D + \frac{J_R}{N})} \tag{8.9}$$

where J_R is the ring current density, J_D is the measured disk current density, and N is the ring collection coefficient (for Pt, ring $N = 0.25$).

8.2.1.2 *OER Analysis*

The OER measurement provides insights into charge reaction kinetics. The important parameter of an OER electrocatalyst is its overpotential (η_{10}) and Tafel slope. The smaller values of overpotential and Tafel slope indicate high catalytic activity and efficiency and can drive the OER more easily and with less energy input which is crucial for rechargeable MABs.

8.2.1.3 *Battery Performance Analysis*

The following key parameters are required to evaluate the performance of rechargeable MABs:

(i) **Capacity** refers to the total amount of charge a battery can store and deliver. It is usually measured in ampere-hours (Ah) or milliampere-hours (mAh) and indicates how long the battery can provide a specific current before the next recharge. Capacity measurement involves fully charging the battery and then discharging it at a constant current until the voltage drops to a specified cutoff level.

(ii) **Energy density** represents the amount of energy that can be stored in a given volume or mass of the battery. It is typically expressed in watt-hours per liter (Wh L^{-1}) or watt-hours per kilogram (Wh kg^{-1}).

(iii) **Cycle life** refers to the number of charge–discharge cycles a battery can undergo before its capacity significantly decreases. It is an essential metric for rechargeable batteries as it indicates their longevity and durability. Cycle-life testing involves repeatedly charging and discharging the battery under controlled conditions until its performance decreases significantly.

8.3 Types of MABs

The MABs have gained intense attention among numerous possibilities. The electrolyte in the system acts as a mediating agent for both the metal anode and the air-breathing cathode. The anode is made up of various metals that have great electrochemical properties, such as alkali metals (Li, Na, K), alkaline earth metals (Mg), and first-row transition metals (Fe, Zn). The type of anode determines whether the electrolyte should be aqueous or nonaqueous. The cathode has a porous and open structure that allows O_2 to flow into the chamber. Different MABs based on different anodes are discussed in the following sections.

8.3.1 *Lithium–Air (Li–O_2) Battery*

A Li–air battery consists of a thin carbon composite as an air cathode, a Li-metal sheet as an anode, and a Li-ion conducting electrolyte. The Li–O_2 batteries have 5–10 times higher energy density than the standard Li-ion batteries. Based on the oxygen mass, the specific-energy density of a Li–O_2 battery is 5200 Wh kg^{-1} or 18.7 MJ kg^{-1}. The maximum theoretical volumetric and gravimetric energy densities in alkaline electrolytes were determined to be 1520 and 1300 Wh kg^{-1}, respectively, while they were found to be 1680 and 1400 Wh kg^{-1} in an acidic environment. The maximum theoretical limits for the specific capacity and energy values mentioned above are solely based on Li metal, air electrode, and electrolyte as active components. Despite possessing greater energy densities compared to standard Li-ion batteries, Li–O_2 batteries initially received less attention. The potential capacity and energy of Li–O_2 batteries are still being debated. In theoretical belief, the cell is constructed in such a manner that the weight and volume of active materials are precisely balanced in the electrochemical reaction. However, some excess electrolytes must be present in practical cells for easy ionic transportation in the entire process of discharge [Zheng *et al.*, 2008]. The types of electrolytes also influence the electrochemistry of Li–O_2 batteries [Girishkumar *et al.*, 2010]. The different types of electrolytes used in Li–O_2 batteries can be classified into four categories, namely aprotic/nonaqueous, aqueous, mixed/hybrid, and solid state. A liquid organic electrolyte is utilized in the aprotic/nonaqueous electrolytic Li–O_2 battery. Li salts such as $LiPF_6$, $LiAsF_6$, $LiN(SO_2CF_3)_2$, and $LiSO_3CF_3$ are frequently used as an electrolyte in organic solvents such as organic carbonates, ethers, and esters [Freunberger *et al.*, 2011]. In the development of Li–O_2 batteries, the nonaqueous system with an organic aprotic electrolyte solution has received significant attention. A Li ion–conducting membrane, using a mixed (aprotic aqueous type) electrolyte, shows successful conduction of Li ions and separates two layers of electrolytes with no moisture impact or cathode blockage. The utilization of polymer, ceramic, or glass electrolytes in solid-state configuration was demonstrated to possess advantageous properties, including the absence of Li dendrite formation, superior thermal stability, and rechargeable capabilities. Despite these positive attributes, solid-state electrolytes exhibit a limiting factor in the form of low ionic conductivity. Some features of Li–O_2 batteries include low volatility, good compatibility of Li anodes, high ionic conduction, and oxidative stability in the Li/Li^+ combination. In Li–O_2 batteries, the investigation was focused on various factors including oxygen solubility, evaporation rates, ionic conductivity, and polarity, as well as different combinations of Li salt and organic carbonate. In 1996, Abraham and Jiang [1996] demonstrated the first true Li–O_2 system with a nonaqueous electrolyte, which included a Li–O_2 battery anode, and proposed two possible energy-producing reactions (8.10, 8.11), as follows:

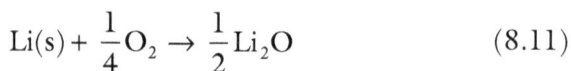

$$Li(s) + \frac{1}{2}O_2 \rightarrow \frac{1}{2}LiO \qquad (8.10)$$

$$Li(s) + \frac{1}{4}O_2 \rightarrow \frac{1}{2}Li_2O \qquad (8.11)$$

The reactions (8.9, 8.10) have a reversible cell voltage (E_o) of 2.959 and 2.913 V, respectively.

Several studies report the utilization of carbon and TMDs as electrocatalysts for air cathode. Girishkumar *et al.* [2010] demonstrated the battery performance using conductive Super P (SP) carbon as a cathode, showing enhanced charge–discharge performance. In this study, they performed a discharge–charge cycle using a Li–O$_2$ cell at a low current density of 0.1 mA cm^{-2} and observed the charging overpotential (η_{chg}) to be higher than discharging overpotential (η_{dis}) with an electrical energy efficiency of 65%. But for the higher efficiency of a system, the overpotential difference should be low. Xia *et al.* [2019] prepared a novel three-dimensional nanostructure, called 1T/2H MoSe$_2$ nanoflowers, using a two-step hydrothermal method. The resulting structure had a large, exposed surface area and exhibits high electrochemical activity, making it an effective bifunctional electrocatalyst for rechargeable Li–O$_2$ batteries. The combination of 1T and 2H phases enhances the synergistic abilities of the material for ORR and OER. In addition, the phase difference facilitates charge transfer kinetics at the interface between the electrolyte and the cathode. The MoSe$_2$ (MS) cathode displays significantly improved coulombic efficiency and charge–discharge capacities compared to the Ketjen black (KB) as cathode. They synthesized 1T/2H MoSe$_2$ nanoflowers at 220°C for 5, 10, 15, and 20 h to obtain MS-5, MS-10, MS-15, and MS-20 samples, respectively. The MS-10 as an electrocatalyst shows a lower oxidation potential for oxygen evolution and a higher potential for oxygen reduction, resulting in a greater peak current density when compared to alternative cathodes. Among the various MS cathode materials, MS-10 demonstrates the superior charge or discharge capacities of 20740 or 21112.47 mAh g^{-1}, respectively, with a remarkable coulombic efficiency of 98% and high cyclic durability up to 100 cycles. The improved performance of the MoSe$_2$ cathode is attributed to its unique nanoflower architecture, which consists of a large number of nanosheets that expose numerous active edge sites.

The current Li–O$_2$ batteries have problems with low capacity and cycle life due to their instability. Li *et al.* [2022] reported a novel sulfur defect engineering approach to enhance the electrochemical performance of Li–O$_2$ batteries by wrapping MoS$_2$ nanoflakes on carbon nanotubes (MoS$_2$@CNT). They utilized TEM to investigate the morphology and microstructure of the sample. The composite structure, shown in Fig 8.3(a), exhibited a fibrous morphology with

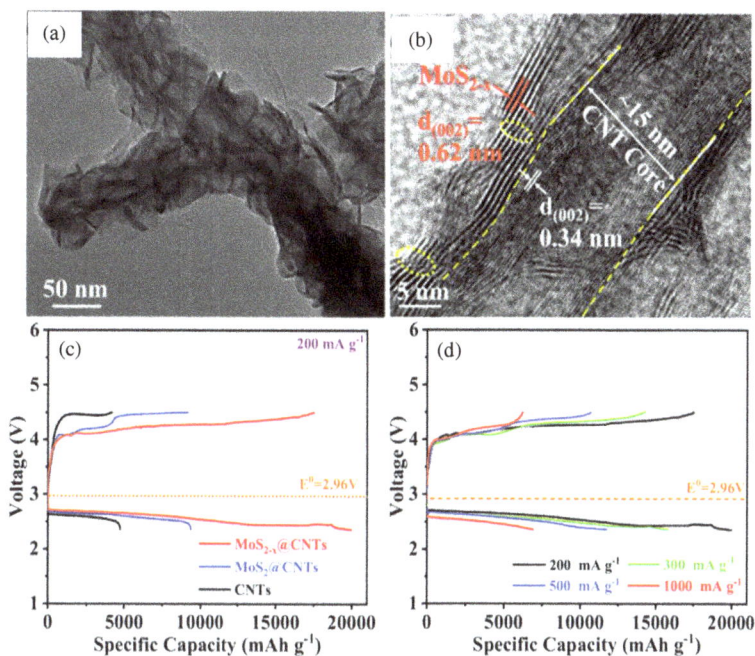

Fig 8.3 (a, b) TEM and HR-TEM image of MoS$_{2-x}$@CNTs, (c) Initial charge–discharge curve from 2.35 to 4.5 V at a current density of 200 mA g^{-1} and (d) rate capability of the MoS$_{2x}$@CNT cathodes, at different current densities Reprinted with permission from Li *et al.* [2022]. Copyright (2023) John Wiley and Sons.

a diameter of about 60 nm and the nanoflakes were evenly distributed and radially wrapped around the CNTs. High-resolution (HR) TEM (Fig 8.3(b)) revealed a core-shell structure consisting of a few layers of thin MoS$_2$ on the exterior and a CNT with a 15 nm diameter on the interior, with an interlayer spacing of MoS$_2$ over CNTs measured as 0.62 nm. The sulfur defect engineering significantly enables the electron transfer between the Li–O$_2$ active material and the carbon surface, thereby enhancing the stability and energy efficiency of the Li–O$_2$ batteries. Fig 8.3(c) shows the specific capacity of different electrodes at the current density of 200 mA g^{-1} within a voltage range of 2.35–4.5 V. The MoS$_{2-x}$@CNT as cathode has the best battery performance, as it has a discharge and charge specific capacity of 19,989 and 17,705 mAh g^{-1} and overpotentials of 0.99 and 0.26 V at 200 mA g^{-1}, respectively. The other electrodes as MoS$_2$@CNT or CNT have less battery performance, as they have lower discharge and charge specific capacities of 9,446 and 9,264 (higher overpotentials of 1.25 and 0.33 V) and 4,802 and 4,227 mAh g^{-1} (higher overpotentials of 1.5 and 0.35 V), respectively. Fig 8.3(d) shows as current density increases, higher overpotential leads to distinct polarization and poorer specific capacities, while the MoS$_{2-x}$@CNT cathode continued to provide better charge–discharge performance.

8.3.2   *Potassium–Air (K–O$_2$) Battery*

The depleting Li resources limit the growth of LIB and Li–O$_2$ batteries. Ren and Wu [2013] fabricated K–O$_2$ batteries to replace Li–O$_2$ batteries due to their low-charging overpotential and good round-trip energy efficiency. The design of the classic K–O$_2$ cell is similar to that of other nonaqueous MABs. The most difficult aspect of building highly durable K–O$_2$ batteries is finding a stable and suitable electrolyte. Owing to its high reducing capacity toward organic electrolytes, the K metal anode requires a solid electrolyte interphase (SEI) to passivate and stabilize it from the by-product K$_2$O$_2$ at the metal anode. The electrolyte must be chemically stable and durable with the superoxide radical anion. Currently, the origin and composition of the SEI on K metal anode remain unclear. The discharge or charge process in K–O$_2$ batteries is initiated by the electrochemical reduction or oxidation of oxygen molecules on the conductive carbon as an air cathode, as shown in the following equations:

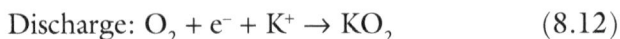

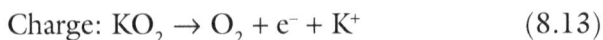

$$\text{Discharge: } O_2 + e^- + K^+ \rightarrow KO_2 \qquad (8.12)$$

$$\text{Charge: } KO_2 \rightarrow O_2 + e^- + K^+ \qquad (8.13)$$

Generated superoxide (O$_2^-$) combines with migrated alkali metal from the anode (K$^+$) and forms metal oxide (KO$_2$). They described a potassium superoxide–based low-overpotential K–O$_2$ battery, which shows that at high charging potential, parasitic reactions between electrolyte and carbon electrode reduce the capacity and shorten the battery life. To examine the effect of cations on the redox chemistry of oxygen, they carried out electrochemical measurements in three electrode setups. Fig 8.4(a) shows the cyclic voltammetry study of ORR for various cations in an aprotic solvent. The electrolyte with K^+ has a much smaller oxygen reduction and oxidation potential gap than the electrolyte with Li^+. A Swagelok $K–O_2$ battery is constructed

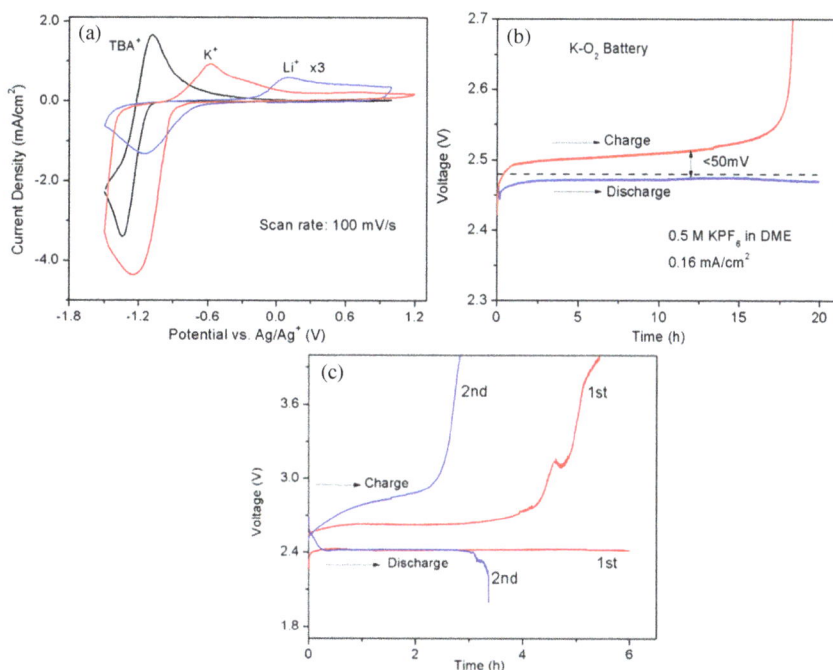

Fig 8.4 (a) Cyclic voltammetry curves for super P–activated carbon on a glassy carbon electrode in three different electrolytes: 0.1 M $LiClO_4$, $TBAPF_6$, or KPF_6, (b) first charge–discharge cycle voltage curves at a current density of 0.16 mA cm^{-2} with electrode geometric area of 0.64 cm^2 for $K–O_2$ battery in 0.5 M KPF_6 in DME as electrolyte and (c) Initial two discharge–charge cycles at 0.16 mA cm^{-2} using 0.5 M KPF6 in butyl diglyme/diglyme mixture with volume ratio 2:5 as electrolyte. Reprinted with permission from Ren and Wu [2013]. Copyright (2023) American Chemical Society.

with a K-metal sheet as anode, a fibrous glassy separator, and a super P–activated carbon-black electrode with a binder on a Ni-foam foundation as an air cathode. A 0.5-M Potassium hexafluorophosphate (KPF_6) as electrolyte was dissolved in an ether solvent (1,2-dimethoxymethane (DME) or diglyme). To compare, a $Li–O_2$ battery was constructed in an exact way using an electrolyte of 1-M Lithium triflate ($LiCF_3SO_3$) in tetraglyme. Fig 8.4(b) displays the voltage plot for the initial charge and discharge of $K–O_2$ and $Li–O_2$ batteries. In the case of $Li–O_2$ battery, the constant discharge plateau region occurs at 2.70 V, resulting in an overpotential of ~260 mV. On the other hand, the constant discharge plateau region is observed at 2.47 V for $K–O_2$ battery, leading to an overpotential of about 10 mV. Due to its higher conductivity (>10 S cm^{-2}, at room temperature) compared with $Li–O_2$, $K–O_2$ battery exhibits a lower discharge overpotential. More crucially, the voltage of $K–O_2$ battery during the subsequent charging potential is as low as 2.50–2.52 V, providing another advantage over $Li–O_2$ battery. Compared to a normal $Li–O_2$ battery with a potential gap of more than 1 V, $K–O_2$ battery has a charge–discharge energy efficacy of >95%. The capacity of $K–O_2$ battery drops with each cycle, but it can still be recharged several times. Fig 8.4(c) illustrates that the first cycle has twice the capacity of the second cycle and the charge voltage is also lower.

It is possible to design a $K–O_2$ battery that uses antimony (Sb) as an anode instead of K metal. It is a silvery-white, brittle metal that is highly resistant to corrosion and is often used in the production of alloys. McCulloch *et al.* [2015] reported that in a $K–O_2$ battery with an Sb anode, the Sb acts as a host material for K ions. When the battery is charged, the K ions are extracted from the cathode and transferred to the Sb anode, where they get stored. The researchers used a K-ion electrolyte and a cathode made of a porous material that allows oxygen to diffuse through it. The Sb anode was prepared by coating the Sb particles onto a conductive substrate through the discharge process. The authors found that the Sb anode had a capacity of ~650 mAh g^{-1} (98% of theoretical values) and demonstrated good stability during repeated charge–discharge cycles. The Sb-composite material was tested for cyclic performance and showed the ability to undergo more than 50 cycles with a capacity of 250 mAh g⁻¹. In addition, the utilization of this anode material was explored in the form of a K_3Sb-O_2 cell, which exhibited remarkable results such as elevated operating voltages, minimal overpotentials, enhanced safety features, and exceptional stability at the interface.

8.3.3 *Sodium–Air (Na–O₂) Battery*

The Na–O_2 battery system advances Li-ion battery technology by replacing Li with sodium (Na), which is abundant in nature. Na interacts with oxygen in Na–O_2 batteries to create sodium superoxide (NaO_2), which is more reactive. The NaO_2 is stable and does not degrade, which helps in the reversal of discharge products during charging [Yin & Fu, 2017]. Na–O_2 battery has a theoretical energy density of 3164 Wh kg⁻¹ and a specific capacity of 1166 mAh g⁻¹, respectively. Rechargeable Li–O_2 batteries have an energy cost of $300–500 kW h⁻¹, whereas Na–O_2 batteries have an energy cost of $100–150 kWh⁻¹, which gives an additional advantage. The Li–O_2 and Na–O_2 batteries with the same electrolyte exhibit distinct oxidation and electrochemical products [Veith *et al.*, 2012]. Na undergoes oxidation and generates Na ions (Na^+) during discharge. The Na^+-conducting NASICON-type ($Na_3Zr_2Si_2PO_{12}$) separator is utilized in hybrid or aqueous Na–O_2 batteries to reduce the blocking of cathode pores with discharge products. Aqueous Na–O_2 batteries undergo the following reactions (8.14–8.18) during their charge–discharge cycle:

During discharge:

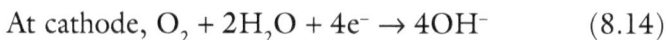

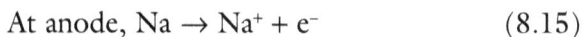

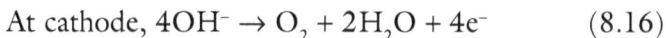

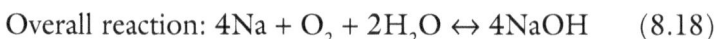

$$At\ cathode,\ O_2 + 2H_2O + 4e^- \rightarrow 4OH^- \qquad (8.14)$$

$$At\ anode,\ Na \rightarrow Na^+ + e^- \qquad (8.15)$$

During charge:

$$At\ cathode,\ 4OH^- \rightarrow O_2 + 2H_2O + 4e^- \qquad (8.16)$$

$$At\ anode,\ Na^+ + e^- \rightarrow Na \qquad (8.17)$$

$$Overall\ reaction:\ 4Na + O_2 + 2H_2O \leftrightarrow 4NaOH \qquad (8.18)$$

In aprotic-based Na–O$_2$ batteries, during discharge and charge, the generated cations flow through the electrolyte, interacting with an e$^-$ from the outer circuit at the cathode and reducing the adsorbed O$_2$ to produce NaO$_2$ which is subsequently reversibly reduced to O$_2$ and Na$^+$. Na–O$_2$ batteries can be aprotic and aqueous or hybrid Na–O$_2$ batteries. In aprotic-based Na–O$_2$ batteries, the nonconductive and insoluble discharge products (Na$_2$O$_2$ and NaO$_2$) are coated or accumulated on a porous air cathode, thereby reducing or blocking the reaction sites and hence the performance of the battery. The reactions (8.19–8.25) for aprotic-based Na–O$_2$ batteries are as follows:
During discharge:

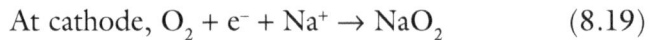

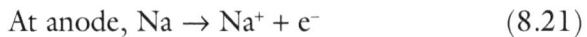

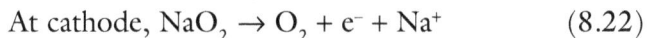

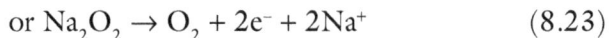

$$\text{At cathode, } O_2 + e^- + Na^+ \rightarrow NaO_2 \qquad (8.19)$$

$$\text{or } O_2 + 2e^- + 2Na^+ \rightarrow Na_2O_2 \qquad (8.20)$$

$$\text{At anode, } Na \rightarrow Na^+ + e^- \qquad (8.21)$$

During charge:

$$\text{At cathode, } NaO_2 \rightarrow O_2 + e^- + Na^+ \qquad (8.22)$$

$$\text{or } Na_2O_2 \rightarrow O_2 + 2e^- + 2Na^+ \qquad (8.23)$$

$$\text{At anode, } Na^+ + e^- \rightarrow Na \qquad (8.24)$$

$$\text{Overall reaction: } 2Na + O_2 \rightarrow Na_2O_2 \qquad (8.25)$$

Peled *et al.* [2011] observed Na–O$_2$ batteries for the first time. They used liquid Na, an air electrode, and a polymer electrolyte to run the Na–O$_2$ battery at 105°C. Sun *et al.* [2012] announced the first room-temperature Na–O$_2$ battery with 20 cycles, a discharge capacity of 1058 mAhg^{-1}, and 85% coulombic efficacy. Liu *et al.* [2013] conducted a study utilizing graphene nanosheets (GNS) as air-electrode catalysts in nonaqueous Na–O$_2$ batteries. Their findings demonstrate that the electrochemical characteristics of the newly designed GNS as air electrodes for Na–O$_2$ batteries exhibit superior performance compared to the traditional carbon electrodes, as illustrated in Fig 8.5(a). The material also shows cyclic charge–discharge performance at a

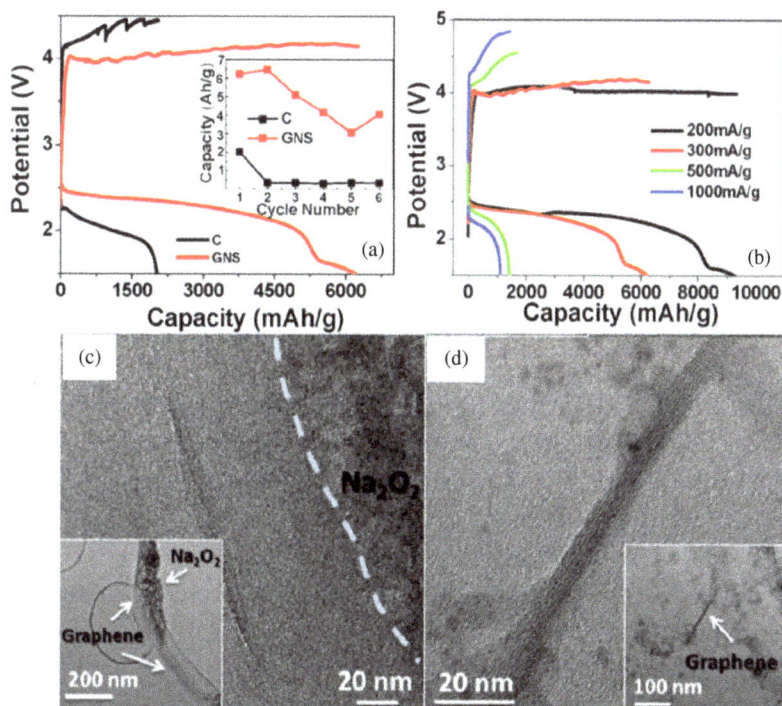

Fig 8.5   (a) Specific capacity charge–discharge voltage curves of Na–O$_2$ battery based on GNS and carbon (their discharge capabilities as a function of cycle number are shown in the inset), (b) GNS–Na cell charge–discharge voltage curves at varying current densities, (c) TEM image of GNS cathode just after the discharging, and (d) TEM image of GNS cathode initial charging of the electrode. Reprinted with permission from Liu *et al.* [2013]. Copyright (2023) Royal Society of Chemistry.

constant current density of 300 mA g^{-1} and reveals that the GNS electrode shows good specific-capacity performance than the carbon electrode. Fig 8.5(b) illustrates that at a current density of 200 mA g^{-1}, the GNS exhibits a high discharge capacity of 9268 mAh g^{-1} compared to decreasing specific capacity of 6208 mAh g^{-1} at 300 mA g^{-1}, 1428 mAh g^{-1} at 500 mA g^{-1}, and 1110 mAh g^{-1} at 1000 mA g^{-1}. The result shows increasing polarization and decreasing specific capacity with increasing current density. They also investigated the structural changes and compositional modification in air electrodes after discharging and charging the electrode. Debris of discharge precipitated product (Na_2O_2) in large amounts and adhere to the GNS can be observed in Fig 8.5(c). The GNS cathode's TEM image pattern after charging is depicted in Fig 8.5(d). On the GNS layers, the majority of the released products vanish.

Murugesan *et al.* [2022] explain the utilization of a low-cost bifunctional electrocatalyst made from a biowaste derivative, a nitrogen-doped highly porous carbon material (N-HPC) for Na–O_2 batteries. The highly porous carbon material was synthesized from biochar, a waste material produced during the pyrolysis of biomass, and was doped with nitrogen to improve its electrocatalytic activity. Porous morphology and many dislocation flaws are produced by structural distortion brought on by the insertion of nitrogen atoms into the graphite lattice.

The authors found that the N-HPC was able to catalyze both the oxygen reduction and evolution reactions at the cathode and anode, respectively, with good performance in a hybrid Na–O_2 battery. The change between charge and discharge voltage (ΔV) value for N-doped HPC was 0.31 V, which suggests better performance than other materials such as sulfur-doped highly porous carbon material (S-HPC) showing 0.41 V. The N-HPC shows high-rate capability at different current densities (6.7–67 mA g^{-1}) and high power density (160 mW g^{-1} at a current density of 70 mA g^{-1}), cycling durability. Overall, the N-HPC as a cathode material-based Na–O_2 battery shows excellent performance such as high bifunctional activity and superior electrical/electronic conductivity, which may be attributed to the mesoporous structure of N-HPC.

8.3.4 *Magnesium–Air (Mg–O_2) Battery*

Mg–O_2 batteries have potential benefits over other MABs, particularly in biomedical and bioelectronics applications. In a neutral electrolyte, Mg–O_2 batteries can potentially provide higher energy density and discharge voltage. The Mg alloy is bioresorbable and the Mg^{2+} ions have no hazardous effect on the human body and the environment. Though,

Mg–O$_2$ batteries have high polarization, lower coulombic efficacy, and substantially lesser operating voltage than theoretically predicted voltage, which restricts their broad utilization during operation. The slow kinetics of the ORR at the air cathode is one of the most significant scientific difficulties that limit the performance of Mg–O$_2$ batteries [Yu *et al.*, 2016]. Reactions (8.26–8.28) are involved in the discharging of the Mg–O$_2$ battery.

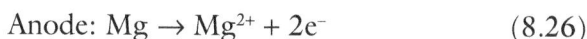

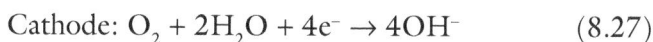

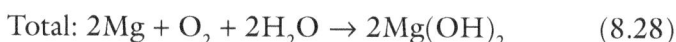

$$\text{Anode: Mg} \rightarrow \text{Mg}^{2+} + 2e^- \tag{8.26}$$

$$\text{Cathode: O}_2 + 2H_2O + 4e^- \rightarrow 4OH^- \tag{8.27}$$

$$\text{Total: 2Mg} + O_2 + 2H_2O \rightarrow 2Mg(OH)_2 \tag{8.28}$$

During the discharge process of electrochemical reaction, O$_2$ flows over the air cathode and undergoes reduction to form hydroxyl ion (OH$^-$), by interaction with water (H$_2$O) and e$^-$ at the opposite electrode, whereas Mg undergoes oxidation and changes to Mg^{2+} with the release of two electrons.

In recent times, Mg–O$_2$ batteries are considered primary batteries. These batteries can be made mechanically reusable or rechargeable by changing the discharged Mg anode and electrolyte with new Mg anode and electrolytes, allowing it to be "refueled." They also have reversible ORR and OER processes and can be recharged electrically. Despite their relatively high voltage and energy density, scientific challenges continue to limit their widespread use. Primary concerns with Mg–O$_2$ batteries are their high polarization and low coulombic efficiency. The deterioration or corrosive nature of the Mg anode induced by the interaction between Mg and the electrolyte, as well as the sluggish or slow kinetics of oxygen reduction at the air cathode, both contribute to drawbacks in Mg–O$_2$ batteries. The other disadvantages of these batteries are the working voltage is often less than 1.2 V and the practical specific-energy density is 10% of the theoretically estimated energy density. Electrolytes in Mg–O$_2$ batteries are typically the neutral saline aqueous solution that comes into contact with the metal electrodes and has a major impact on the electrode reactions. The electrolyte and the electrodes in the Mg-O$_2$ battery

Table 8.1 Corrosion potential of "bare" Mg in different aqueous solutions.

Electrolyte	Rest Potential (V *vs* NHE)
NaCl	–1.72
Na_2SO_4	–1.75
HCl	–1.68
HNO_3	–1.49
NaOH	–1.47
NH_3	–1.43

Source: [Zhang *et al.*, 2014]

cause high polarization and poor coulombic efficiency. Thus, selecting an appropriate electrolyte is critical for the battery's performance. The corrosion potential of "bare" Mg was studied by Zhang *et al.* [2014] in different aqueous solutions, as summarized in Table 8.1. The Mg has stronger corrosion resistance in the alkaline medium than in acidic or neutral media, as seen in Table 8.1.

The current study also demonstrates that raising the pH above 10 can suppress the HER process and improve the Mg corrosion resistance by using an electrolyte made up of a nearly saturated aqueous solution of $MgCl_2$, LiCl, or their combination. The strong corrosion resistance of magnesium in alkaline solution is attributed to the partial formation of an $Mg(OH)_2$ coating on the surface of Mg alloy or pure magnesium. This may help to prevent the corrosion of anode-active material. However, excessive $Mg(OH)_2$ layer formation on the electrode stops the anode from reacting further, resulting in a delayed response which leads to load increase. Therefore, in Mg–O_2 batteries, a neutral electrolyte is often employed [Winther-Jensen *et al.*, 2008; Zhang *et al.*, 2014]. Interestingly, the rechargeable Mg–O_2 battery has a theoretically estimated energy capacity of 3.9 kWh kg^{-1} and volumetric density of 14 kW hL^{-1} and both these values are higher than the theoretical values for Li–O_2 cell. For the scientific use of this battery, high-efficiency catalysts are urgently needed to improve the reactivity of MgO with nonaqueous electrolytes.

Cheng *et al.* [2018] demonstrated that to overcome the slow reaction kinetics of air cathodes in Mg–O_2 batteries, an open-meso-porous carbon nanofiber (CNF) with homogeneously connected atomic Fe–N_x sites (OM-NCNF-FeN_x) as air catalyst was incorporated in Mg–O_2 batteries. They used a polished Mg sheet/foil as anode, a carbon cloth loaded with OM-NCNF-FeN_x as the catalyst for the air cathode, and Dulbecco's phosphate-buffered saline (DPBS) medium as the electrolyte, as shown in Fig 8.6(a). Fig 8.6(b) shows SEM images and schematic pictures of OM-NCNF-FeN_x ink (right) and Pt/C ink (left) loaded on the conductive carbon cloth. This innovative oxygen electrode's remarkable electrocatalytic and battery performance can be due to the following factors:

1) ORR's activity and stability were enhanced by the homogeneous coupling of atomic Fe–N_x sites.
2) The active Fe–N_x sites were fully accessible because of the open-mesoporous and linked structures, as well as the large specific surface area, which boost the mass transport qualities.
3) The air diffusion paths were significantly increased by the three-dimensional hierarchically porous channels and networks in the cathode.

In neutral electrolytes (DPBS), using OM-NCNF-FeN_x in Mg–O_2 batteries can function efficiently with high open-circuit voltage (OCV), constant discharge voltage plateaus, high capacity, and long operation life, which is better than novel Pt/C equipped batteries, as shown in Fig 8.6(c-f) [Cheng *et al.*, 2018]. As a result, the obstacles and high demands for effectual air cathodes in Mg–O_2 batteries with neutral electrolytes may be addressed by this innovative and high-performance air electrode. Moreover, this study opens up new possibilities for the large-scale production of various nanofibrous carbon electrodes for various applications, including future MAB design to achieve flexible, wearable, and bio-adaptable power sources for various electronic and biomedical devices.

Li *et al.* [2016] have reported the effectiveness of combined ultra-fine Mn_3O_4 nanowires, three-dimensional graphene, and single-walled

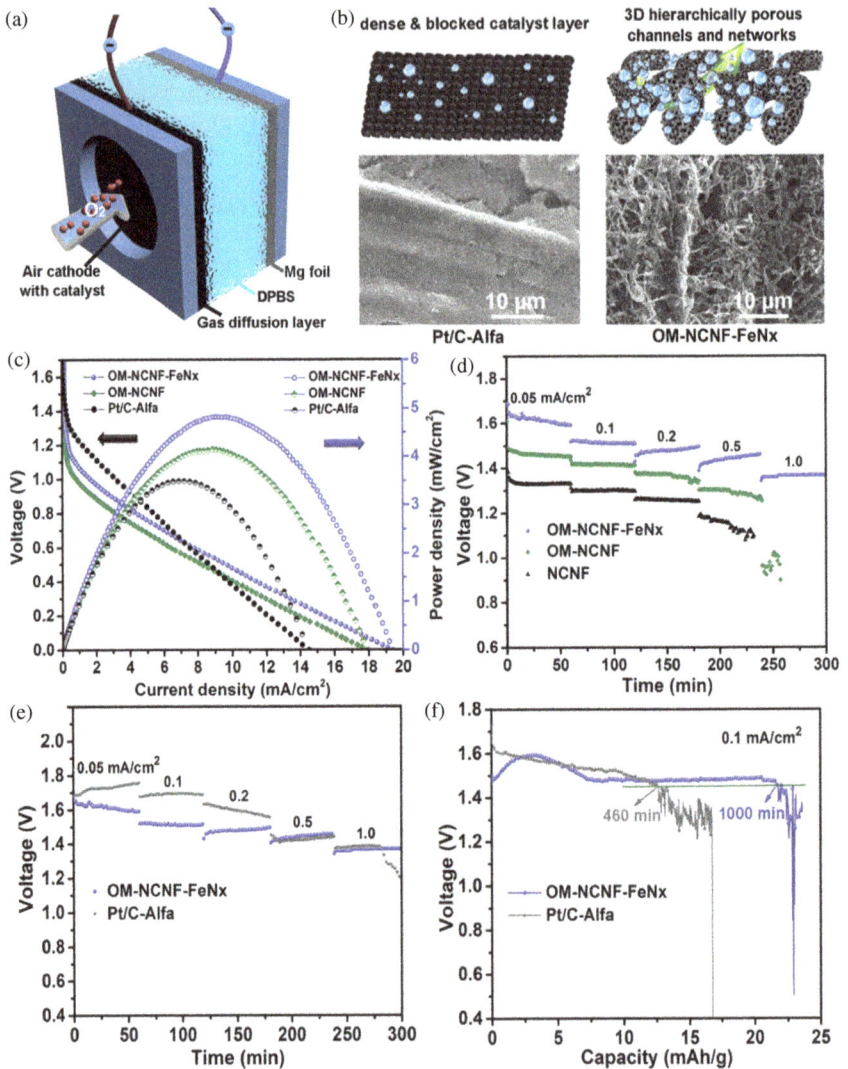

Fig 8.6 (a) Schematic representation of the aqueous Mg–air battery, (b) Pt/C ink on carbon cloth (left), OM-NCNF-FeN$_x$ ink on carbon cloth (right), and SEM pictures and diagrams (right), (c) discharge voltage and power density curves of primary liquid Mg–air batteries in DPBS with Pt/C-Alfa and different CNF catalysts, (d) galvanostatic discharge curves of batteries in DPBS electrolyte with various CNF catalysts at different current densities, 60 min each, (e) discharge curves, and (f) 0.1-mAcm^{-2} current-density discharge capacity of liquid Mg–air batteries with OM-NCNF-FeN$_x$ and Pt/C-Alfa cathodes in DPBS. Reprinted with permission from Cheng *et al.* [2018]. Copyright (2023) John Wiley and Sons.

CNT to create composites that exhibit exceptional electrocatalysts activity for ORR and enhanced performance for $Mg-O_2$ batteries. This study concluded that morphology changes in Mn_3O_4 nanostructures have a significant effect on the electrochemical performance of $Mg-O_2$ batteries. The $Mg-O_2$ based on the reported electrocatalyst shows a higher OCP of 1.49 V, a voltage plateau of 1.34 V, and a longer service time of 4177 min.

8.3.5 *Zinc–Air (Zn–O₂) Battery*

$Zn-O_2$ batteries can be considered as both fuel cells and batteries. In $Zn-O_2$ batteries, the Zn metal acts as fuel and catalytic reaction can be controlled by varying the controlled flow of air. Prior studies elucidate that Zn and iron in metallic form are more stable and robust, have higher cell voltage, and higher energy in a liquid or aqueous electrolyte, and can be charged more efficiently in aqueous electrolytes. Zn is inexpensive and abundant in nature compared to Li. Specific energy density of Zn metal (1353 Wh kg^{-1}) is less compared to Li metal (5928 Wh kg^{-1}) and other MABs [Fu *et al.*, 2017; Zhang *et al.*, 2019].

During discharge:

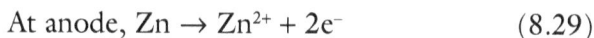

$$\text{At anode, } Zn \rightarrow Zn^{2+} + 2e^- \qquad (8.29)$$

Electrons are liberated at the Zn anode during oxidation reaction (8.29)

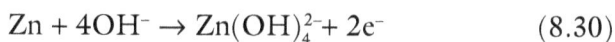

$$Zn + 4OH^- \rightarrow Zn(OH)_4^{2-} + 2e^- \qquad (8.30)$$

During the discharge process, metallic Zn anode in contact with the alkaline electrolyte generates zincate ion (reaction 8.30)

$$Zn(OH)_4^{2-} \rightarrow ZnO + H_2O + 2OH^- \qquad (8.31)$$

Generated zincate ion decomposes to form solid ZnO powder (reaction 8.31).

$$\text{At cathode, } O_2 + 2H_2O + 4e^- \rightarrow 4OH^- \qquad (8.32)$$

$$O_2 + H_2O + 2e^- \rightarrow HO_2^- + OH^- \qquad (8.33)$$

$$HO_2^- + H_2O + 2e^- \rightarrow 3OH^- \qquad (8.34)$$

$$2HO_2^- \rightarrow 2OH^- + O_2 \qquad (8.35)$$

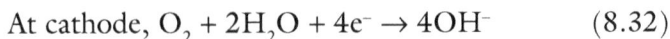

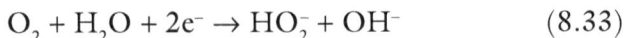

Two types of pathways $4e^-$ or $2e^-$ occur at the metal catalyst surface embedded in air cathode during ORR in aqueous electrolyte media as described in the above-mentioned reactions. The $4e^-$ pathway is the direct reaction and it is more favorable than the $2e^-$ pathway due to its enhanced energy efficiency. The $2e^-$ reaction pathway unfavorably affects the formation of peroxide species which causes corrosiveness and leads to untimely performance degradation of the Zn–O_2 battery [Fu *et al.*, 2017]. Simultaneously, diffused oxygen molecules into porous air electrodes are reduced into hydroxyl ion (OH^-) in ORR (reaction 8.32), which migrates from one end of the cathode to the other end of the anode. There are other possibilities as well that follow the $2e^-$ reaction-pathway reduction of peroxide (reaction 8.33) and chemical disproportionation of peroxide (reactions 8.34 and 8.35). The overall reaction is mentioned in reaction 8.36.

$$\text{Overall reaction: } 2Zn + O_2 \rightarrow 2ZnO \qquad (8.36)$$

Prabu *et al.* [2014] successfully synthesized $CoMn_2O_4$ (CMO) using a one-pot hydrothermal synthesis technique and then decorated it onto surfaces of rGO that were either nitrogen-doped or undoped. The resulting hybrid material, CMO/N-rGO, showed exceptional charge–discharge performance, along with extended cycle stability, when used as a bifunctional air electrode in primary and rechargeable Zn–O_2 batteries, even in high concentrations of aqueous alkaline electrolyte under ambient air conditions. This study specifically investigates the rechargeability and durability of bifunctional air electrodes in real-world circumstances, including open-air environments. SEM images in Fig 8.7(a) and (b) depicted the wrinkled hybrid structures of CMO/rGO and CMO/N-rGO, respectively. The stability of the

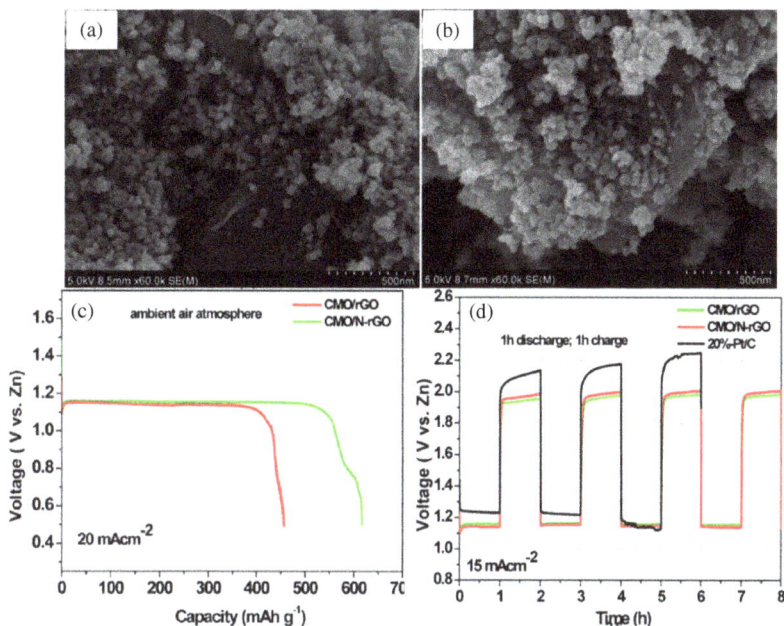

Fig 8.7 (a, b) SEM images, (c) discharge curve of CMO/rGO and CMO/N-rGO at a current density of 20 mA cm^{-2}, and (d) long-run charge–discharge performance of 2 h per cycle at a current density of 15 mA cm^{-2} for CMO/NrGO, CMO/rGO, and Pt/C. Reprinted with permission from Prabu *et al.* [2014]. Copyright (2023) American Chemical Society.

catalyst under deep discharge conditions was evaluated at a current density of 20 mA cm^{-2}. The hybrid electrocatalyst of CMO/N-rGO demonstrated an OCV of ~1.5 V, followed by a stable potential plateau lasting for 12 h at 1.15 V, outperforming the CMO/rGO for about 16 h. In addition, the discharge capacity of CMO/N-rGO, the hybrid electrocatalyst, was significantly higher at 610 mAh g^{-1} compared to the specific capacity of 460 mA g^{-1} exhibited by the CMO/rGO cathode, as illustrated in Fig 8.7(c). The superior performance of CMO/N-rGO in terms of OCV, plateau duration, discharge time, and discharge capacity can be attributed to the N-doping in rGO, which enhances the number of charge carriers, as well as the morphological effects of pyridinic and graphitic-nitrogen.

In addition, the incorporation of nitrogen into the graphene layer induces the formation of defects at the edges, leading to increased adsorption and retention of oxygen molecules at these sites, thereby enhancing the catalytic performance of the $Zn-O_2$ battery during extensive discharge. Fig 8.7(d) shows the assessment of a rechargeable $Zn-O_2$ battery charged and discharged at a current density of 15 mA cm^{-2} with 2-h intervals. Authors analyzed the charge–discharge performance of different electrocatalysts including CMO/N-rGO, CMO/rGO, and Pt/C. Their results revealed that the use of a hybrid-bifunctional air electrode made of CMO and N-rGO led to improved charge–discharge performance. The CMO/N-rGO showed an initial charge and discharge overpotential of 0.51 and 0.33 V, respectively, which slightly increased to 0.53 and 0.35 V after the eighth cycle. On the other hand, Pt/C displayed a rapid surge in charge and discharge overpotential, starting from 0.41 and 0.55 V to reaching 0.61 and 1.04 V after the eighth cycle. Overall, due to their exceptional structure, morphology, and electrocatalytic activity, the CMO/N-rGO hybrid materials exhibit great potential as a superior material for $Zn-O_2$ battery applications, outperforming commercial Pt/C catalysts.

Maurya *et al.* [2023] created a new, hybrid nanostructure electrocatalyst made of $MoSe_2$ and $NiCo_2O_4/NiO$ to be used in rechargeable $Zn-O_2$ batteries. It was reported to have improved performance compared to existing ones and showed enhanced efficiency and cyclic longevity of $Zn-O_2$ batteries. SEM (Fig 8.8(a)) image shows the wrinkle few layers of formation, whereas TEM (Fig 8.8(b)) shows the lattice fringes of the $MoSe_2$ nanosheet. In Fig 8.8(c), SEM shows the nanocluster formation of $NiCo_2O_4/NiO$, and TEM (Fig 8.8(d)) shows the formation of nanocluster with different sizes.

SEM (Fig 8.8(e)) image of $NiCo_2O_4/NiO-MoSe_2$ shows the hybrid nanostructure and reveals the uniform distribution of interconnected wrinkly $MoSe_2$ nanosheets and $NiCo_2O_4/NiO$ nanoclusters. The conjoint nature of $MoSe_2$ and $NiCo_2O_4/NiO$ in the hybrid nanostructure is revealed by TEM (Fig 8.8(f)) image, signifying the homogeneity of the hybrid nanostructure. The $NiCo_2O_4/NiO-MoSe_2$ hybrid nanostructure as a cathode in $Zn-O_2$ showed the best performance, with OCV of 1.42 V than $NiCo_2O_4/NiO-$ and $MoSe_2$-based

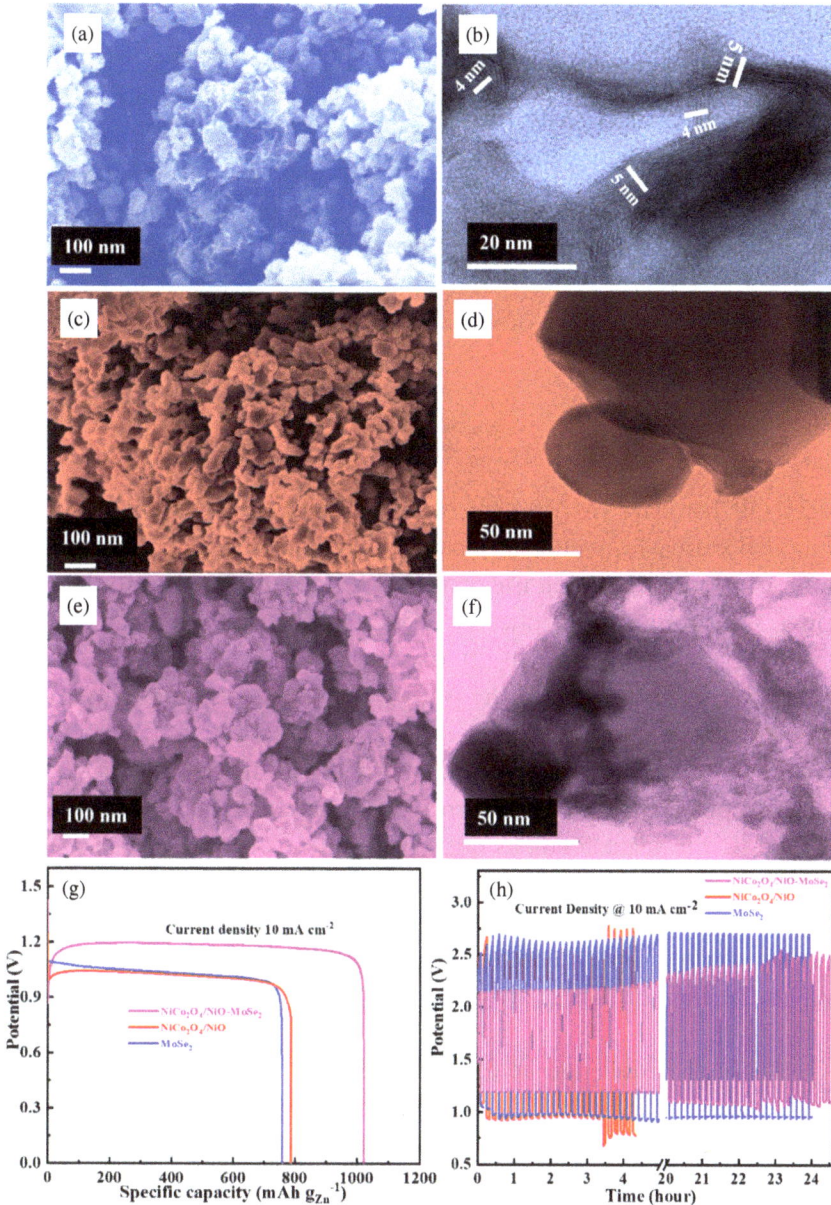

Fig 8.8 SEM and TEM images of (a, b) $MoSe_2$, (c, d) $NiCO_2O_4/NiO$, and (d, e) hybrid $NiCo_2O_4/NiO\text{-}MoSe_2$ nanostructure. (f) Specific capacity at a current density of 10 mA cm^{-2} and (g) charging–discharge curves at a current density of 10 mA cm^{-2} of $MoSe_2$, $NiCo_2O_4/NiO$, and hybrid $NiCo_2O_4/NiO\text{-}MoSe_2$ nanostructure. Reprinted with permission from Maurya *et al.* [2023]. Copyright (2023) Elsevier.

Zn–O$_2$ battery. The hybrid nanostructure showed a high specific capacity of 1023 mAh g$_{Zn}^{-1}$ compared to 760 mAh g$_{Zn}^{-1}$ for MoSe$_2$ and 788 mAh g$_{Zn}^{-1}$ for NiCo$_2$O$_4$/NiO, as shown in Fig 8.8(g). The MoSe$_2$ in hybrid nanostructure contributes to the durability and high performance of zinc–air batteries. This is because MoSe$_2$ has good stability, which means it can withstand external changes and can maintain its structure without breaking down easily. The stability of MoSe$_2$ helps to ensure that the hybrid nanostructure remains intact, thus enhancing the overall performance of the Zn–O$_2$ batteries as shown in Fig 8.8(h).

8.4 Summary, Challenges, and Prospects

Metal–air batteries (MABs) are becoming more popular due to their ability to store a high energy density, which positions them as a promising solution for meeting the growing demand for energy storage and conversion in modern applications such as electric vehicles and smart grids. Recent advancements in material selection, preparation, and processing techniques have the potential to enhance the performance of MABs for next-generation energy devices. Bifunctional electrocatalysts made of low-cost materials, such as doped carbon or TMD-based nanocomposites, have been extensively researched and demonstrated. Currently, the main focus in the development of MABs is on manufacturing air-electrodes that serve as bifunctional electrocatalysts for both the ORR and OER. However, there are a number of obstacles that should be overcome in order to successfully bring MABs technology to the market. There are a few challenges as follows.

(i) Lack of knowledge of electrocatalyst's interaction and synergy processes with other components on a fundamental/basic level.

(ii) Existing electrocatalysts have insufficient bifunctional (ORR and OER) activity and stability for air-electrode.

(iii) Battery performance is degraded due to corrosion of electrodes.

(iv) Undesirable side reactions between electrolyte and electrocatalysts cause a reduction in stability and cycle life.

To overcome the constraints due to the limited functioning of air electrodes, researchers are investigating the capability of nonmetallic (heteroatom)-doped carbon materials and TMD-based nanocomposites as effective electrocatalysts for MABs. Corrosion and oxidation of electrodes degrade their performance. Innovative methods need to be developed for making high-performance, bifunctional electrocatalysts from low-cost, sustainable, and renewable biomasses containing heteroatoms such as O, K, N, S, Mg, or Ca. Researchers have to find out how to make these types of bifunctional catalysts derived from biomass as air-electrodes for MABs. In addition, issues such as biomass material optimization, carbon corrosion, and the relationship between nitrogen doping and battery performance, among others, have to be addressed using innovative technologies and methodologies. Controlling and reducing undesirable electrolyte and air-electrode side reactions is also needed.

References

Abraham, K. M. & Jiang, Z. (1996). A polymer electrolyte-based rechargeable lithium/oxygen battery. *J. Electrochem. Soc.* 143(1), pp. 1–5.

Antoine, O. & Durand, R. (2000). RRDE study of oxygen reduction on Pt nanoparticles inside Nafion: H_2O_2 production in PEMFC cathode conditions. *J. Appl. Electrochem.* 30(7), pp. 839–844.

Chang, Z., Xu, J., Liu, Q., Li, L. & Zhang, X. (2015). Recent progress on stability enhancement for cathode in rechargeable non-aqueous lithium-oxygen battery. *Adv. Energy Mater.* 5(21), p. 1500633.

Chau, K. T., Wong, Y. S. & Chan, C. C. (1999). An overview of energy sources for electric vehicles. *Energy Convers. Manag.* 40(10), pp. 1021–1039.

Cheng, F. & Chen, J. (2012). Metal–air batteries: From oxygen reduction electrochemistry to cathode catalysts. *Chem. Soc. Rev.* 41(6), pp. 2172–2192.

Cheng, C., Li, S., Xia, Y., Ma, L., Nie, C., Roth, C., Thomas, A. & Haag, R. (2018). Atomic Fe-N_x coupled open-mesoporous carbon nanofibers for efficient and bioadaptable oxygen electrode in Mg–air batteries. *Adv. Mater.* 30(40), p. 1802669.

Dunn, B., Kamath, H. & Tarascon, J.-M. (2011). Electrical energy storage for the grid: A battery of choices. *Science* 334(6058), pp. 928–935.

Freunberger, S. A., Chen, Y., Drewett, N. E., Hardwick, L. J., Bardé, F. & Bruce, P. G. (2011). The lithium–oxygen battery with ether-based electrolytes. *Angew. Chem. Int. Ed.* 50(37), pp. 8609–8613.

Fu, J., Cano, Z. P., Park, M. G., Yu, A., Fowler, M. & Chen, Z. (2017). Electrically rechargeable zinc–air batteries: Progress, challenges perspectives. *Adv. Mater.* 29(7), p. 1604685.

Ge, X., Goh, F. W. T., Li, B., Hor, T. S. A., Zhang, J., Xiao, P., Wang, X., Zong, Y. & Liu, Z. (2015). Efficient and durable oxygen reduction and evolution of a hydrothermally synthesized $La(Co_{0.55}Mn_{0.45})_{0.99}O_{3-\delta}$ nanorod/graphene hybrid in alkaline media. *Nanoscale* 7(19), pp. 9046–9054.

Girishkumar, G., McCloskey, B., Luntz, A. C., Swanson, S. & Wilcke, W. (2010). Lithium–air battery: promise and challenges. *J. Phys. Chem. Lett.* 1(14), pp. 2193–2203.

Huang, H., Feng, X., Du, C. & Song, W. (2015). High-quality phosphorus-doped MoS_2 ultrathin nanosheets with amenable ORR catalytic activity. *Chem. Commun.* 51(37), pp. 7903–7906.

Li, Y. & Lu, J. (2017). Metal–air batteries: Will they be the future electrochemical energy storage device of choice? *ACS Energy Lett.* 2(6), pp. 1370–1377.

Li, Y., Zhang, X., Li, H. B., Yoo, H. D., Chi, X., An, Q., Liu, J., Yu, M., Wang, W. & Yao, Y. (2016). Mixed-phase mullite electrocatalyst for pH-neutral oxygen reduction in magnesium-air batteries. *Nano Energy*. 27, pp. 8–16.

Li, D., Zhao, L., Xia, Q., Wang, J., Liu, X., Xu, H. & Chou, S. (2022). Activating MoS_2 Nanoflakes via Sulfur defect engineering wrapped on CNTs for stable and efficient $Li-O_2$ batteries. *Adv. Fun. Mater.* 32(8), p. 2108153.

Liu, W., Sun, Q., Yang, Y., Xie, J.-Y. & Fu, Z.-W. (2013). An enhanced electrochemical performance of a sodium-air battery with graphene nanosheets as air electrode catalysts †. *Chem. Commun.* 49, p. 1951.

Ma, Z., Wang, K., Qiu, Y., Liu, X., Cao, C., Feng, Y. & Hu, P. A. (2018). Nitrogen and sulfur co-doped porous carbon derived from bio-waste as a promising electrocatalyst for zinc-air battery. *Energy* 143, pp. 43–55.

Maurya, P. K., Mishra, S. & Mishra, A. K. (2023). $MoSe_2$ and $NiCo_2O_4/$ NiO based hybrid nanostructure as novel electrocatalyst for high performance rechargeable zinc-air battery. *Electrochimica Acta* 439, p. 141689.

McCulloch, W. D., Ren, X., Yu, M., Huang, Z. & Wu, Y. (2015). Potassium-ion oxygen battery based on a high capacity antimony anode. *ACS Appl. Mater. Interfaces* 7(47), pp. 26158–26166.

Murugesan, C., Senthilkumar, B. & Barpanda, P. (2022). Biowaste-Derived highly porous N-Doped carbon as a low-cost bifunctional electrocatalyst for hybrid sodium-air batteries. *ACS Sustain. Chem. Eng.* 10(28), pp. 9077–9086.

Peled, E., Golodnitsky, D., Mazor, H., Goor, M. & Avshalomov, S. (2011). Parameter analysis of a practical lithium- and sodium-air electric vehicle battery. *J. Power Sources* 196(16), pp. 6835–6840.

Prabu, M., Ramakrishnan, P., Nara, H., Momma, T., Osaka, T. & Shanmugam, S. (2014). Zinc-air battery: Understanding the structure and morphology changes of graphene-supported $CoMn_2O_4$ Bifunctional catalysts under practical rechargeable conditions. *ACS Appl. Mater. Interfaces* 6(19), pp. 16545–16555.

Rao, Z. & Wang, S. (2011). A review of power battery thermal energy management. *Renewable and Sustainable Energy Rev.* 15(9), pp. 4554–4571.

Ren, X. & Wu, Y. (2013). A low-overpotential potassium-oxygen battery based on potassium superoxide. *J. Am. Chem. Soc.* 135(8), pp. 2923–2926.

Sharifi, T., Hu, G., Jia, X. & Wågberg, T. (2012). Formation of active sites for oxygen reduction reactions by transformation of nitrogen functionalities in nitrogen-doped carbon nanotubes. *ACS Nano* 6(10), pp. 8904–8912.

Shinagawa, T., Garcia-Esparza, A. T. & Takanabe, K. (2015). Insight on Tafel slopes from a microkinetic analysis of aqueous electrocatalysis for energy conversion. *Sci. Rep.* 5(1), p. 13801.

Sun, Q., Yang, Y. & Fu, Z. W. (2012). Electrochemical properties of room temperature sodium–air batteries with non-aqueous electrolyte. *Electrochem Commun.* 16(1), pp. 22–25.

Taniguchi, A., Fujioka, N., Ikoma, M. & Ohta, A. (2001). Development of nickel/metal-hydride batteries for EVs and HEVs. *J. Power Sources* 100(1–2), pp. 117–124.

Veith, G. M., Nanda, J., Delmau, L. H. & Dudney, N. J. (2012). Influence of lithium salts on the discharge chemistry of Li–air cells. *J. Phys. Chem. Lett.* 3(10), pp. 1242–1247.

Wang, C., Zhang, Z., Abedinia, O. & Farkoush, S. G. (2021). Modeling and analysis of a microgrid considering the uncertainty in renewable energy resources, energy storage systems and demand management in electrical retail market. *J. Energy Storage* 33, p. 102111.

Winther-Jensen, B., Gaadingwe, M., Macfarlane, D. R. & Forsyth, M. (2008). Control of magnesium interfacial reactions in aqueous electrolytes towards a biocompatible battery. *Electrochimica Acta* 53(20), pp. 5881–5884.

Wu, G., Cui, G., Li, D., Shen, P. K. & Li, N. (2009). Carbon-supported Co1.67Te2 nanoparticles as electrocatalysts for oxygen reduction reaction in alkaline electrolyte. *J. Mater. Chem.* 19(36), pp. 6581–6589.

Xia, Q., Zhao, L., Li, D., Wang, J., Liu, L., Hou, C., Liu, X., Xu, H., Dang, F. & Zhang, J. (2021). Phase modulation of 1T/2H $MoSe_2$ nanoflowers for highly efficient bifunctional electrocatalysis in rechargeable Li-O_2 batteries. *J. Mater. Chem. A* 9(35), pp. 19922–19931.

Yin, W. W. & Fu, Z. W. (2017). The Potential of Na-air batteries. *ChemCatChem* 9(9), pp. 1545–1553.

Yu, C., Wang, C., Liu, X., Jia, X., Naficy, S., Shu, K., Forsyth, M., Wallace, G. G., Yu, C., Wang, C., Liu, X., Jia, X., Naficy, S., Shu, K., Wallace, G. G. & Forsyth, M. (2016). A cytocompatible robust hybrid conducting polymer hydrogel for use in a magnesium battery. *Adv. Mater.* 28(42), pp. 9349–9355.

Zhang, T., Tao, Z. & Chen, J. (2014). Magnesium-air batteries: From principle to application. *Mater. Horiz.* 1(2), pp. 196–206.

Zhang, X., Wang, X. G., Xie, Z. & Zhou, Z. (2016). Recent progress in rechargeable alkali metal–air batteries. *Green Energy Environ.* 1(1), pp. 4–17.

Zhang, J., Zhou, Q., Tang, Y., Zhang, L. & Li, Y. (2019). Zinc–air batteries: Are they ready for prime time?. *Chem. Sci.* 10(39), pp. 8924–8929.

Zheng, J. P., Liang, R. Y., Hendrickson, M. & Plichta, E. J. (2008). Theoretical energy density of li–air batteries. *J. Electrochem. Soc.* 155(6), p. A432.

Chapter 9

Carbon-Based and TMDs-Based Nanostructured Anode Materials for Improved Lithium-Ion and Sodium-Ion Battery Performances

Vimal K. Tiwari[1] and Rajendra Kumar Singh[2]

[1]*Department of Physical Sciences, Banasthali Vidyapith,
Banasthali Vidyapith, Banasthali 304022, India*
[2]*Ionic Liquid and Solid-State Ionics Lab, Department of Physics,
Institute of Science, Banaras Hindu University,
Varanasi 221005, India*
Emails: vimalkumartiwari@banasthali.in; vimalitbhu@gmail.com

Abstract

The hierarchical designing and structural modifications of nano-structured materials for optimum battery performance have been exploited to fulfill the growing demand of high energy-storage devices. Precise synthesis methods, doping, and structural modification of active materials have shown great advantages in restricting phase transformation, volume expansion, dendritic growth, poor ionic and electronic conductivity, transition metal dissolution, etc. Nanostructured materials (e.g., carbon and transition metal dichalcogenides (TMDs)) have significant scientific and industrial importance owing to the higher surface area to volume ratio, molecular

259

level interactions, and possible appearance of quantum effects at the nanoscale. This chapter focuses on the role of nanostructured materials as anode materials (negative electrodes) for rechargeable battery applications and assists the design of safe, high capacity, long cyclability, and high energy density cell formation with carbon and TMDs electrode materials.

9.1 Introduction

The increasing global energy requirement is mainly fulfilled by fossil fuel and nuclear energy sources and is shifting toward environmental friendly renewable energy sources. Fossil fuels are finite and produce unfriendly gases such as greenhouse gases and carbon substances responsible for global warming. Moreover, nuclear energy sources emit radioactive substances and nuclear waste. Green and renewable energy sources, including solar, wind, and hydropower, are good alternatives to limited fossil energy in terms of environmental pollution. However, their discontinuous energy production and other shortcomings do not serve continuous, long-time energy requirements. To address these concerns, cost-effective energy-storage systems (ESSs) are of great importance for continuous and portable energy requirements and are used to integrate the scattered energy as large-scale ESSs. Different types of battery technologies are needed to fulfill the growing demands of energy-storage devices in many applications, for example, daily use appliances, portable devices, stationary storage grids, and transport systems. The rechargeable lithium-ion battery has proved to be an efficient potable battery device and is used in different types of advanced electronic gadgets. Each cell of the battery undergoes electrochemical reactions through the movement of ions between the cathode and anode, which have different electrochemical potentials. Both electrodes control the redox reactions and play an important role in obtaining high electrochemical performance of the battery. Recent concerns about galvanic cells defined as a battery are mainly material-related issues that limit energy density, lead to high polarization, and restrict it for high-performance battery applications. During cycling, the active materials are forced to undergo volume and structural changes responsible for fast capacity fading, resulting in low electrochemical performance [Fu *et al.*, 2006]. In addition, dendrites formation, uneven electrodeposition of anode ions, and delamination of foil current collectors are responsible for the decrease in capacity and cyclability of the battery. Batteries fail to perform at high temperatures and in humid conditions which create safety concerns at extreme environmental conditions. Therefore, a large number of studies have focused on building up stable anode and cathode materials with high electrochemical performance. An atomic layer electrode material can fulfill the required demand of storage capacity. The microstructure containing nanometer-sized particles provide an easy ion transport pathway which leads to lower ion mobility barriers (ultrafast ion transport) and potentially allows the high current density with outstanding mechanical and chemical stabilities. The high stress- and strain-bearing abilities of nanostructured materials allow them to be used as mechanically supportive high-capacity electrode materials in battery applications. Due to altered structural and physical properties, their electrical conductivity can be tuned for cathode, anode, or separator. These nanostructured materials, such as graphene, graphene analogs, and chalcogenides 2D materials, also show corrosion-protective property and are potential candidates to handle environmental sustainability and safety issues in battery applications. These nanostructured materials used as anode improve thermal safety, mechanical stability, cyclability, and volumetric and gravimetric capacities; prevent dendrite formation; and avoid pulverization. This chapter discusses the recent advances and perspectives of nanostructured particles, including carbonaceous, its nanohybrids and transition metal dichalcogenides (TMDs) as anode materials for lithium-ion battery (LIB) and sodium-ion battery (SIB) applications.

9.2 Role of Anode in the Battery

A battery is a combination of cells which are connected according to the requirement, in series and parallel, whereas the cell components are a cathode, an anode, and an electrolyte. During charging/discharging, the battery experiences electrical current through the movement of ions by a chemical reaction between anode and cathode through electrolyte and electron movement by an external circuit.

In charging, the movement of positive ions (e.g., Li^+ or Na^+) is from cathode to anode, whereas, during discharging, it is reversed (for LIBs or SIBs). Lithium (Li) metal (theoretical specific capacity ~3860 mAh g^{-1}) is as an anode in LIBs. Li as a negative electrode (anode) was first suggested by Hajek in 1949. Li as an anode for LIBs and Na metal as an anode for SIBs were first considered in 1980 by the researchers. The output voltages of a single-cell LIB and NIB are ~3 and 2.8 V, respectively, which are much higher than lead–acid batteries, NiCd, and NiMH cells. Moreover, Li/Na batteries have many advantages, such as requiring less cells for a given voltage, higher energy density, and low self-discharge. The LIB-based storage devices are predominately used in portable electronics and electronic vehicles and are also tested in aircrafts and many other places. The increase in price and the diminishing accessibility of Li metal will restrict its use in the large-scale energy-storage grid and will be a serious concern in the future for all LIBs energy storage–based devices. To date, SIBs are fitted as analogs of LIBs due to low cost and abundant Na resources. Na has similar physical and chemical properties as Li, and SIB follows a similar electrochemical storage mechanism as LIB. The low-cost components of SIBs, including Na as an anode, aluminum (Al) as a current collector in place of copper, and electrolytes, make it more attractive to researchers due to the huge demand of energy-storage devices and limited resources of Li. However, the corrosion and dendrite formation on Li/Na metal anode during battery cycling performance can cause short circuits between the anode and the cathode by penetrating through the separator. Therefore, highly reactive Li/Na metal anode has safety concerns during charging/discharging. A successful alternative approach to solve the problem related to Li/Na metal has been implemented to supplant anode with other materials that store Li/Na ions through intercalation/deintercalation at a low voltage during charging/discharging, respectively. Generally, the anodes for LIBs and SIBs are carbon-based materials, and cathodes are made of transition metal oxides (TMOs) and polyanions. The charge/discharge reactions on the anode and cathode and transportation of Li^+ ions in LIBs are mentioned through reactions as follows [Ali *et al.*, 2015].

9.2.1 *During Charging*

At the cathode, Li^+ ions are released and transported through electrolytes in the internal circuit, whereas electrons move through the external circuit to anode.

$$LiMO_2 \rightarrow Li_{1-x}MO_2 + xLi^+ + xe^- \qquad (9.1)$$

At the anode, both Li^+ and electrons are restored. M denotes transition metal and C carbon material.

$$xLi^+ + xe^- + C \rightarrow Li_xC \qquad (9.2)$$

9.2.2 *During Discharging*

At anode, Li^+ ions are released and transported through electrolytes in internal circuit, whereas electrons move through external circuit to cathode.

$$Li_xC \rightarrow xLi^+ + xe^- + C \qquad (9.3)$$

And at cathode, both Li^+ and electrons are restored.

$$Li_{1-x}MO_2 + xLi^+ + xe^- \rightarrow LiMO_2 \qquad (9.4)$$

The above reaction shows flow of Li^+ ion through the electrolyte between anode and cathode during charge and discharge. Similar electrochemistry is obtained when Na^+ ions are used in SIBs.

9.3 Types of Anode Materials for LIB and SIB Applications

Owing to high energy density, good retention, and environmental safety, the rechargeable LIBs and SIBs have been established as suitable ESS to provide energy in all formats, from portable to grid level. The Li anode is extremely reactive due to the low potential of reduction (−3.04 V versus standard hydrogen electrode) and the continuous charge/discharge cycle leads to dendrite formation on the

Li anode, eventually resulting in a short circuit during cycling. On the contrary, carbon-based graphite anode shows a lower theoretical capacity (372 mAh g^{-1}) with an intercalation potential of 0–0.3 V vs Li/Li$^+$. The graphite-based carbon material replaced Li metal as an anode in 1983, where Li ions move through intercalated and deintercalated in LIBs during cycling, and the schematic representation is shown in Fig 9.1(a) [Yazami & Touzain, 1983]. The low-cost, carbon-based anode materials minimize dendrite formation with good stability during charge/discharge cycles. Carbon materials such as graphite show 372 mAh g^{-1} theoretical capacity by forming LiC$_6$.

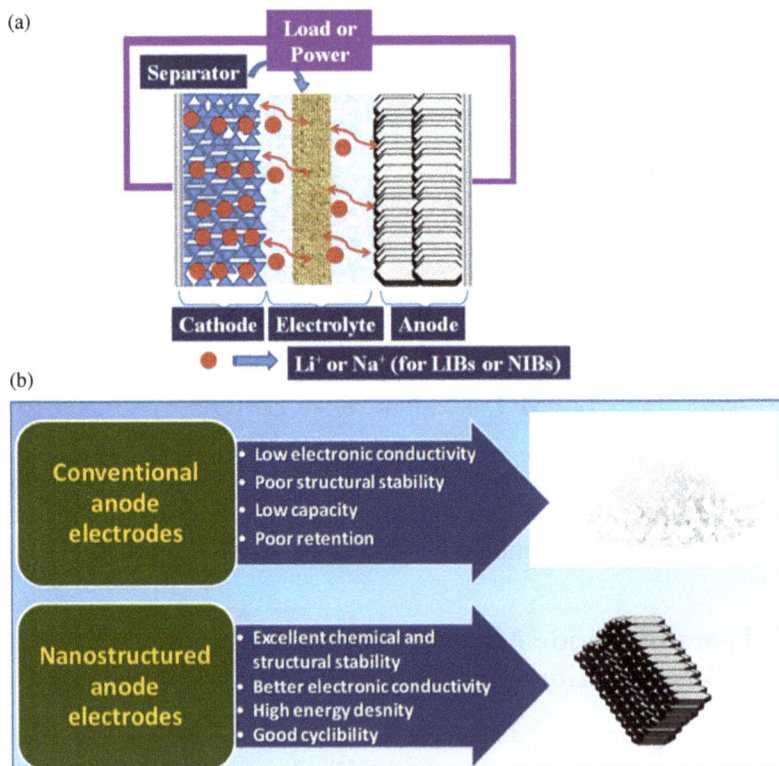

Fig 9.1 (a) Schematic representation of cell performance, which is common for both LIBs and SIBs and (b) the disadvantages and advantages of conventional and nanostructured anodes are mentioned, respectively.

The limited capacity of these nonmetal anodes restricts the generation of required energy and power density for battery application. In addition, the emerging high-capacity alloys (silicon (Si) and germanium (Ge)) and transition metal compounds anodes have poor electrical conductivity, volume expansion, significant capacity loss, and pulverization and cannot resolve electrochemical instability issues [Yan *et al.*, 2019]. The microparticles show relatively low power density due to less surface area and high polarization caused by moderate diffusion of ions and slow electron conduction. The use of nanostructured material electrodes (such as graphene, CNTs, TMD), instead of usual electrodes, provides higher Li storage capacity due to high surface area which offers easy diffusion for ions and electron transportation, more suitable for battery application. Many studies on LIBs confirm that the performance of anode materials is improved by decreasing the particle size, which accommodates high stress and strain without structural deformation.

SIB is established as an alternative promising candidate due to its safer and lower cost than LIB, especially for grid-level energy-storage applications. Still, the poor cyclability and low energy density of SIB need to be optimized, and continuous investigation is required for designing grid-level energy-storage applications. Advanced nanomaterials synthesized and designed as anode material would significantly improve electron and ion transportation, enhancing the electrochemical properties of battery systems. There are many studies on various compounds for SIBs as cathode materials, similar to cathodes of LIBs, including layered oxides, polyanionic, Na superionic conductor (NASICON) structure phosphates, tunnel oxides compounds, etc. The high-capacity, layered-oxide $NaCoO_2$ cathode material for SIB, similar to $LiCoO_2$ for LIB, which forms different types of phase structures, was first time reported by Wang *et al.* [2017].

In general, cathode materials for SIBs have similar compositions and similar phase structures like the cathodes of LIB. On the contrary, however, graphite and many other common anode materials used for LIBs are not appropriate for Na intercalation in SIBs due to larger size of Na and unfavorable thermodynamics. The main concerns for SIB anode materials are their large volume expansion, moderate rate performance, and low coulombic efficiency. Different types of nanostructured anode materials are reported for SIBs which are soft and hard-carbon materials, TMDs, titanium-based alloy, and organic materials.

9.4 Types of Nanostructured Materials as Anode

Several nanostructured materials, including carbon, Si, TMDs, and oxide materials, can be tuned for their electrical properties according to requirements in different applications [Yan *et al.*, 2019]. Specially designed nanostructured materials (e.g., carbon and TMDs) as battery anodes have been successfully implemented, which play an important role in facile electron and ion conduction due to different atomic architectures such as puckered, planar, and buckling structures. The usage of Si, TMOs, and transition metal hydroxides (TMHs) nanostructured materials as anode reduces capacity decay for battery application [Fang *et al.*, 2020]. In addition, the 2D-layered materials (TMO/TMH) demonstrate abundant intercalation sites during Li^+/Na^+ intercalation and provide short diffusion paths, resulting in high discharge capacity, small volume change, and a fast charging rate during cycling. The disadvantages and advantages of conventional and nanostructured anode materials are highlighted in Fig 9.1(b). The most common approaches such as mechanical exfoliation and chemical vapor deposition (CVD) to synthesize nanostructured materials have some limitations, for example, inhomogeneous thickness issues, oxidation, and low yield. In contrast, the solution route emerges as a powerful synthetic technique to develop versatile nanostructures, including quantum dots, 1D and 2D nanostructured materials with large surface area for widespread production and applications. However, use of high melting metal precursors is a drawback which limits the hierarchical growth of nanostructures at an ambient condition. The nanostructured-based anode materials are classified into three types of materials: nanostructured carboneous, TMDs, and other materials for LIBs and SIBs.

9.4.1 *Nanostructured Carbonaceous Materials*

The high surface area and electronic conductivity of anode materials are keys to achieving high-performance electrochemical properties. The conducting-reduced graphene oxide (RGO), CNTs, and other nanostructured carbon as anode have been studied in detail to replace graphite for LIBs. Graphene and CNTs are commonly used as anode materials due to good electronic conductivity, high charge carrier mobility, and Li storage capacity as compared to graphite for battery applications. The 1D carbon nanofibers (CNFs) with outstanding electronic conductivity are used in cathode materials as conductive additives. CNFs also have excellent Li-storage performance when used as anode in LIBs. Owing to electronic, ionic, conducting and porous structures, nanostructured carbonaceous materials facilitate the movement of charge carriers to improve the electrochemical performance. The environmentally friendly, low-cost TMD materials are used to prepare anode, which shows high theoretical capacities, although improvements are still required for poor rate performance and low capacity retention. The nanostructured hybrid of metal oxide and nanostructured carbon can enhance the Li storage properties with improved electronic and ionic charge transfer kinetics and mechanical strength during the battery performance. The precursors for carbonaceous materials are pure graphite, hydrocarbons, resin, and pitch etc. [Nan *et al.*, 2014]. The carbon anode materials made of high-cost resin-based precursors show high performance, whereas anodes based on hydrocarbon precursors with acceptable raw material value produce low carbon yield. The cheapest anthracite with the highest carbon yield shows lower storage capacity compared to hard-carbon materials. The nanostructured design is also an important factor for anode material along with the carbon yield and price to select the appropriate carbonaceous materials.

9.4.1.1 *Carboneous Materials as Anode for LIBs*

The carbon-based nanostructured materials as anode for LIBs consist of various morphologies such as graphite, graphene, RGO, CNTs, carbon nanorods, carbon xerogel (CX), and CNFs. These are discussed in detail in the following sections.

9.4.1.1.1 CNTs-based anode materials

CNTs are composed of rolled graphene sheets with strong C–C covalent bonds and show high electronic conductivity, inert nature, and excellent mechanical strength [Landi *et al.*, 2009]. They are classified as single-walled CNTs (SWCNTs) and multiwalled CNTs (MWCNTs) depending on their structure. The specific surface area of SWCNT of diameter ~1–2 nm is 2630 $m^2 g^{-1}$. An anode made of graphene ink for LIB shows a high capacity of ~165 mAh g^{-1} and ~190 Wh kg^{-1} energy density at 1C rate over 80 cycles [Hassoun *et al.*, 2014]. Further, the rolled, concentric cylinders of graphene sheets named as MWCNTs show higher capacity, good conduction properties, and ballistic electron transport [Bachtold *et al.*, 1999]. The SWNTs deliver a specific capacity of 450 mAh g^{-1} at 74 mA g^{-1} as compared to theoretical specific capacity (372 mAh g^{-1}).

The Li-ion capacity of MWCNT anode synthesized by CVD demonstrates excellent reversibility and coulombic efficiency at a low current with reference to a Li-metal electrode. A reversible capacity for the MWCNT paper is measured higher than 225 mAh g^{-1} at 74 mA g^{-1} after 20 cycles when using the electrolyte EC:PC:DEC. The Li-ion capacity of MWCNT paper is improved to 340 mAh g^{-1} after the change in catalyst solvent from xylenes to pyridine and shows a stable reversible capacity during cycling [Landi *et al.*, 2009]. The battery performance of CNTs depends on the preparation method due to structural defects, impurities present, and graphitization.

9.4.1.1.2 CNFs-based anode materials

One-dimensional sp^2-hybridized CNFs with high electronic conductivity provide path to electrolytes, resulting in improved Li-storage performance. CNFs have different structural configuration from CNTs and can be prepared by various methods such as catalytic CVD growth, electrospinning, and template-based synthesis. CNFs are cost-effective anodic materials for LIBs due to highly graphitized synthesis at low temperatures, which is suitable for bulk production. The highly graphitized CNFs were synthesized by CVD in the range from 550°C to 700°C, which could be designed from platelet (P) structure to tubular (T). The synthesized CNFs anode exhibits specific capacity of ~297–431 mAh g^{-1} for LIB [Yoon *et al.*, 2004]. Moreover, a suitable doping in CNFs anode materials enhances the battery performance. The randomly oriented and interconnected morphology of nitrogen-doped CNFs (NCNFs) shows a porous and nonwoven network of hundreds of nanometers in diameter, as shown in Fig 9.2(a). The freestanding structure of NCNFs indicates good

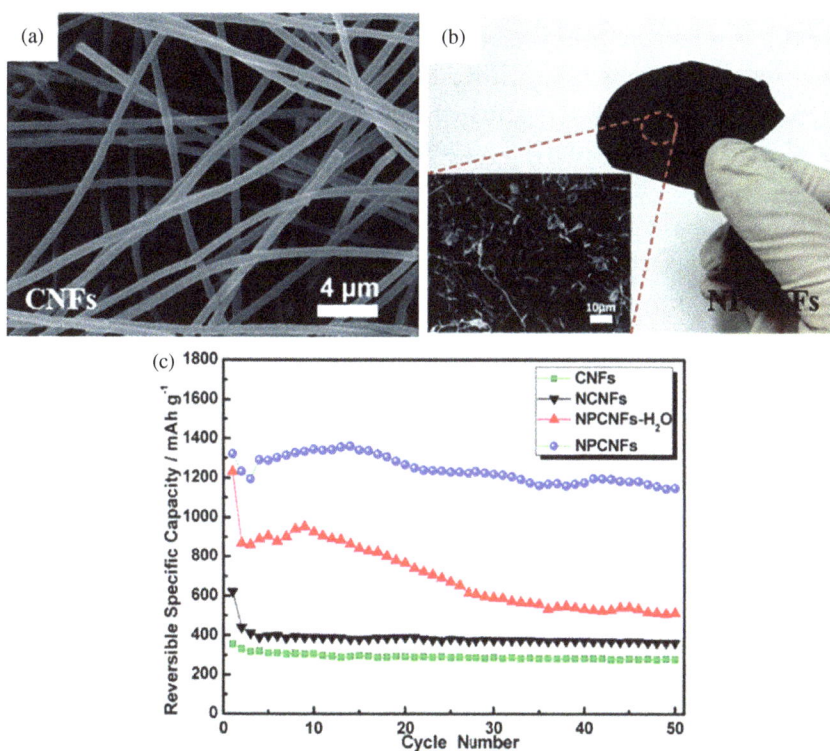

Fig 9.2   (a) CNFs networking are shown by SEM image, (b) sheet of freestanding NPCNFs (inset is the SEM image of that NPCNFs) and (c) electrochemical cycling performance for different composition of CNFs. Reprinted with permission from Nan *et al.* [2014]. Copyright (2023) Royal Society of Chemistry.

mechanical integrity, as shown in Fig 9.2(b). The cycling performance of nitrogen-doped porous CNF anodes demonstrates high capacity and good cycling stability with an initial discharge capacity of 1323 mAh g^{-1} at the current 50 mA g^{-1} (Fig 9.2(c)). The flexible CNF doped with nitrogen or oxygen shows better initial specific capacities and capacity retention of ~702 mAhg^{-1} at 5 A g^{-1} after 500 cycles [Wang *et al.*, 2016].

9.4.1.1.3 CX-based anode materials

The porous CXs are synthesized using copolymerization of hydroxylated benzene with aldehyde [Canal-Rodríguez *et al.*, 2018]. CX is a continuous nanoporous structure made of active C and graphite with low density. Porous CXs with a high degree of graphitization were prepared by microwave heating of the pristine CX. The improvement in electrical conductivity was ~90%. The synthesized graphitized anode showed 70% higher specific capacities compared to pristine CX with stable cycle performances of LIBs. The CX–SiO composite made of graphite, SiO, Si crystals, and active C showed a uniform dispersion of CX and improved discharge capacity [Yuan *et al.*, 2007]. The above promising anode composites materials show high specific capacity for LIBs in the range of 0–1.5 V.

9.4.1.1.4 Graphene-based anode materials

Yoo *et al.* [2008] first investigated graphene sheets as anode materials through the reduction of GO by hydrazine for LIBs. RGO anode exhibited an initial capacity of 540 mAh g^{-1} with a fast decay in capacity. Further, heat treatment of graphene sheets at 500°C showed an improved capacity of 650 mAh g^{-1} with good cycling stability [Wang *et al.*, 2009]. Whereas Nan *et al.* [2014] reported very high reversible capacity (>1000 mAh g^{-1}) with thermally reduced graphene sheets anode materials. The graphene-based anode materials showed a high storage capacity of >800 mAh g^{-1} with high battery performance [Wang *et al.*, 2009]. The elevated storage capacity of these materials could be due to increased surface area, larger interlayer design, and reduced diameter. The stretchable anodes were obtained by incorporating dopants in graphene structures. Aerogels, CNT foam, and fiber cloth made of graphene were used as anode for elastic stretchable anode material for self-supported LIB [Canal-Rodríguez *et al.*, 2018; Yuan *et al.*, 2007]. The composite of TMOs and RGO also helped to enhance specific capacity, energy density, and cycle life for LIB [Lv *et al.*, 2015].

9.4.1.1.5 Carbon hybrids anode materials

A composite anode made of SnO_2 nanoparticles covered with graphene nanosheets (GNSs) showed improved cyclability with specific capacity of 570 mAh g^{-1} after 30 cycles compared to 60 mAh g^{-1} of bare SnO_2 nanoparticles anode after 15 cycles at 50 mA g^{-1} current density [Paek *et al.*, 2009]. The mechanical failure of these composite anodes restricts them from practical applications. The binder-free electrode made of graphene-wrapping of Si discontinues the pulverization of Si nanoparticles and delivers a capacity of 878 mAh g^{-1} compared to 200 mAh g^{-1} of Si electrode over 100 cycles at 100 mA g^{-1} current density [Wang *et al.*, 2015]. The conversion transition metal compounds anodes (e.g., Fe_3O_4 with high theoretical capacity ~924 mAh g^{-1}), which are low-cost and ecofriendly materials, undergo fast reaction kinetics and result in high specific capacity during reversible redox reactions. However, these types of anode materials experience capacity loss due to electrode pulverization and large-volume expansion. To accommodate changes in volume during cycling and improve battery performance, nanostructured carbon materials have widely been used as mechanically strong high-conducting host and manage volume expansion of Fe_3O_4 nanoparticles upon lithiation. The graphene-encapsulated Fe_3O_4 nanosphere networks (Fe_3O_4@GS/GF) and grafted mesoporous Fe_3O_4 nanoparticles on 3D graphene foams resist aggregation and volume changes and improve electrochemical performance with low Li$^+$ ion-transfer resistance [Wei *et al.*, 2013]. Carbon-coated nanostructured hematite (Fe_2O_3) anode material synthesized by carbonization of ferrocene precursor shows good capacity ~1138 mAhg^{-1} after 300 cycles at current density 0.5 A g^{-1}

and still retains capacity of ~458.8 mAh g^{-1} at 10 A g^{-1} for LIB. The abundant, nontoxic hematite (α-Fe$_2$O$_3$) anode material has high theoretical capacity (1007 mAh g^{-1}), which shows high corrosion resistance properties. However, some drawbacks include fast capacity fading and poor high-rate performance during cycling. The carbon-coated iron-oxide anode material that avoids the fast capacity fading due to carbon covering shows good electrochemical performance with a stable and wide voltage window [Zhang *et al.*, 2008]. The nanostructured carbon-coated α-Fe$_2$O$_3$ anode material has been synthesized by using ferrocene precursor via simple pyrolysis to prepare hematite for LIBs and delivers a high initial capacity of 1138 mAh g^{-1} after 300 cycles at 500 mA g^{-1}. The CNT-encapsulated 1D Sn nanowire (SnNW@CNT) hybrid nanomaterial as an anode is evaluated through molecular dynamics (MD) and DFT calculations for LIBs [Ng *et al.*, 2010]. An optimized SnNW@CNT is found to be thermodynamically stable and can store more Li$^+$ as compared to pure CNT due to a nanostructured void space during cycling performance. The ultrathin SnNW uptake more Li due to favorable alloying process and large topological defects (nonagonal ring), providing easy path for ion diffusion.

9.4.1.2 *Carbon-based Anode Materials for SIBs*

The well-established graphite-based anode material for LIBs shows lower capacity in SIBs due to insufficient interplanar spacing between layers for Na$^+$ insertion. In contrast, the disordered carbon anode material with defect sites and nanovoids exhibits improved reversible capacity in SIB. The full cell SIBs, including anthracite-derived anode and Na$_{0.9}$[Cu$_{0.22}$Fe$_{0.30}$Mn$_{0.48}$]O$_2$ cathode materials, have been successfully configured as promising candidates and are shown to have cycling stability and good rate performance with the energy density of ~100 Wh kg^{-1} [Li *et al.*, 2016]. The drawback of hard carbon is the large capacity contribution at a low potential plateau ~0.2 V (vs. Na$^+$/Na), causing the Na plating, which results in safety concerns. The low-capacity soft-carbon anode materials show stable operation voltage. A comprehensive study has been done on types of carbon material-based anodes of SIBs.

9.4.1.2.1 Modified graphite-based anode materials

A well-established graphite anode for LIBs is inappropriate for Na intercalation in its layered structure for SIBs. Modified graphite with large interplanar spacing provides a suitable large interlayer space for Na intercalation during cycling. This modified graphite displays high reversible capacity (284 mAh g^{-1}) compare to pristine graphite (~35 mAh g^{-1}) with good cycling performance. But, modified graphite with low conductivity suffers a low coulombic efficiency [Ge & Fouletier, 1988]. The RGO, a highly conducting anode material, gives a good shape charge–discharge curve with a higher reversible capacity of ~450 mAh g^{-1} at 25 mAh g^{-1}, stable rate performance, and high cycle performance due to high surface area for Na adsorption as shown in Fig 9.3(a-c) [Wan *et al.*, 2016].

Further, these anode materials show a high electrochemical performance for SIBs by surface defects through nitrogen-doping in

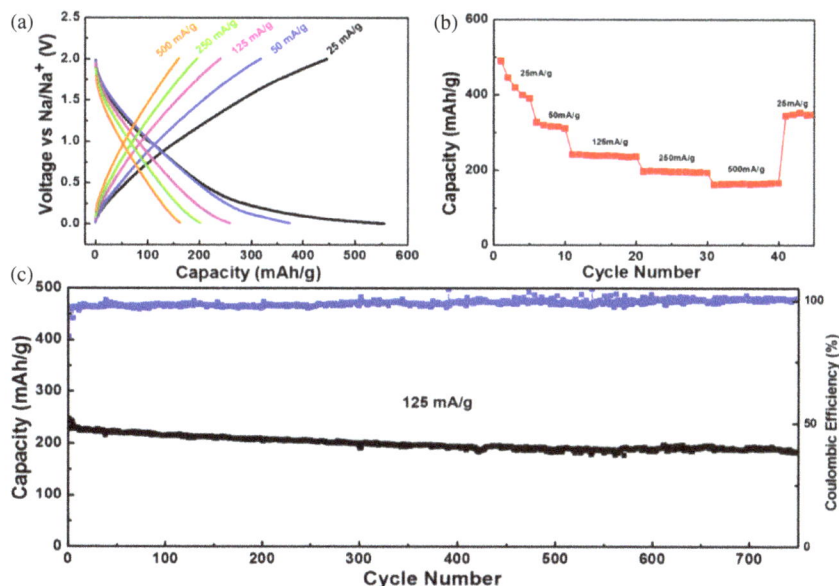

Fig 9.3.   SIB performance of rapid-RGO anode (a) charge discharge plot at 500, 250, 125, 50, and 25 mA g^{-1}, (b) rate performance at different current rates, and (c) cyclability performance >700 cycles. Reprinted with permission from Wan *et al.* [2016]. Copyright (2023) American Chemical Society.

3D graphene foams [Wang *et al.*, 2017]. Another study with phosphorene-conducting graphene hybrid material found that hybrid material can achieve a discharge capacity of 2440 mAh g^{-1} with 83% retention over 100 cycles at 50 mA g^{-1} (0.02 C) in the presence of conducting graphene for SIBs [Sun *et al.*, 2015]. The thin GNS as anode material also demonstrates high electrochemical Na-storage properties due to its superior conducting nature and high surface area. The CNF thin film with nitrogen and boron co-doping appears to have high electronic conductivity and reactivity due to defects, and this modified anode material illustrates a higher capacity of 581 mAh g^{-1} at 100 mA g^{-1} current rate. In addition, a hierarchical nanocomposite of porous carbon with few wt% loading of highly conducting GNSs, which reduce Na ions diffusion path, shows better cycling stability and a high specific capacity of 670 mAhg^{-1} at 50 mAg^{-1}. Hierarchically porous structures and high-conductivity graphene additives effectively enhance the rate/cycle performance. Moreover, theoretically, it is also established that 3D interconnected porous carbon structure has shown better structure stability than heteroatom doping and can provide easy Na-ions transportation for long-cycle performance [Wang *et al.*, 2017].

9.4.1.2.2 Soft carbon materials

The soft carbon materials, nongraphite in nature, are used as anodes for SIBs and show less twisted and regular graphitic layers with high operation potential compared to hard carbon. The most aromatic compounds such as pitch, petroleum coke, plastics, and tar are used as precursors to synthesize soft carbon. The soft carbon anode material synthesized by petroleum coke first reported the specific capacity of ~90 mAh g^{-1} for SIBs [Wang *et al.*, 2017]. Whereas, an anthracite derived based soft-carbon anode material shows a high capacity of ~222 mAh g^{-1}. The defects are created in soft carbon by heteroatom doping, that is, nitrogen, sulfur, and phosphorous, in structure to get high electronic conductivity and high specific surface area which provide higher reversible capacity and higher sodiation potential in SIBs.

9.4.1.2.3 Hard-carbon materials

The hard-carbon materials synthesized at high temperatures are micropores rigid structure and amorphous in nature, are cross-linked with aligned graphitic layers, and give space for Na intercalation. These anode materials show a low potential plateau through insertion into amorphous micropores and a sloping potential region due to interaction with defect sites of Na during charge–discharge process [Wang *et al.*, 2017]. Hard-carbon materials can be obtained by pyrolysis to get controlled microstructure and more defect sites of solid-phase precursors including sugars, peels from apples and peanuts, polymers, and wood [Wang *et al.*, 2017]. The different processes and pretreatment methods, such as gas flow rate, temperature, and time, strongly impact the structure and properties of hard carbon. The hard-carbon anode material synthesized from the pitch-phenolic resin demonstrates a specific capacity of 284 mAhg^{-1} at a low current rate with high coulombic efficiency (~88%) and good cycling performance. A hard carbon prepared at 1150°C from polyaniline precursor shows a high capacity of ~270 mAhg^{-1} at 50 mAg^{-1} [Wang *et al.*, 2017]. The monodispersed hard-carbon spherule anode material was found from sucrose precursor by carbonization to 1600°C, illustrating a high capacity of ~220 mAh g^{-1} and ~93% retention over 100 cycles at low current rate. Hard-carbon microtubes synthesized by cotton precursor and carbonized at different temperatures, 1000°C and 1300°C, delivered high capacities of 315 mAh g^{-1} and 88 mAh g^{-1}, respectively. The hard-carbon anode with 17% expanded spacing in the graphene interlayer due to high defective sites synthesized from banana peels showed a high specific capacity of ~355 mAh g^{-1}, superior cyclability, and stable coulombic efficiency [Wang *et al.*, 2017].

9.4.2 *TMD-based Anode Materials*

9.4.2.1 *TMD-based Anode Materials for LIBs*

Graphite, currently used as commercial anode with low theoretical specific capacity of 372 mAh g^{-1} and low operating voltage, has limited applications for LIBs [Yazami and Touzain, 1983; Yoo *et al.*, 2008].

To overcome the limitations of graphite, new electrodes should be low-cost, durable, and environmentally friendly to be used as commercial anode. The semiconducting and insulating TMD nano-materials, such as MoS_2, WS_2, $MoSe_2$, and WSe_2, with similar structural properties to semi-metallic graphene, have tunable electrical, mechanical, and optical properties for electronic devices, optoelectronic devices gas sensing, and energy-storage devices. Recently, TMDs, the 2D-layered materials, have been established as anode materials due to their remarkable electrochemical properties providing large surface area between interatomic layers and accommodating volume expansion for LIBs. Due to the limited practical performance of the pristine carbon materials with theoretical capacity of 372 mAh g^{-1}, the TMDs with CNTs are implemented as next-generation anode candidates with high electrochemical performance for LIBs. TMDs have received considerable attention on energy-storage applications due to unusual electrical conductivities with unique physical and chemical properties. In TMDs, different MX_2 layers made of one layer of transition metal (M) atoms covered with two layers of chalcogen (X) are attracted by weak van der Waals forces. The WS_2 has large interatomic spacing of 0.62 nm showing diffusion energy barrier, which is beneficial to improve the intralayer conductivity [Srinivaas *et al.*, 2019]. The increase in interplanner spacing of MoS_2, WS_2, and other TMDs helps to promote carrier diffusion kinetics and superior electronic conductivities during Li-ion transportation. The enlarged space supports Li-ion intercalation and reduces the barrier to enhance ion mobility. However, the cycling stability is affected by a rapid decay in capacity due to low conductivity and volume expansion of TMDs during the cycling process, resulting in decrease in the number of active sites. To solve these problems, TMD-based composites were prepared with electrical-conducting carbon materials. A hierarchical structure of the TMD anode has been prepared to increase Li-ion and electron transportation. The two distinct symmetries such as the trigonal prismatic 2H phase (D3h) and the octahedral 1T phase (Oh) of the WS_2 layer lie at the position of the S atoms [Srinivaas *et al.*, 2019]. The low electrical-conducting 2H phase of WS_2 shows a high volume expansion during the electrochemical performance whereas, the 1T phase has good electron conduction and better interaction with electrolytes to maximize the utilization of the active material. Therefore, the formation of pure metallic 1T phase is an efficient method to achieve good electrical-conducting TMDs showing improved electrochemical performance. However, synthesizing the pure metallic 1T phase of WS_2 remains challenging. Tungsten ditelluride (WTe_2), a type-II Weyl semi-metal, shows a positive QSH gap in the monolayers and displays high electrical conductivity. On the other hand, the morphology of the as-synthesized few-layered WS_2 nanoflowers (NFs) demonstrates porous structure as shown in the SEM images (Fig 9.4(a-c)). Further, the chemical composition analysis by EDS mapping of WS_2 NFs confirms the presence of W and S, as shown in Fig 9.4(d,e). WTe_2 (WTe_2@CNT) nanocomposite delivers a high specific capacity of 1097 mAh g^{-1} at 100 mA g^{-1}, whereas WTe_2 nanostars show a reversible capacity of 655 mAh g^{-1} at the same current of LIB [Srinivaas *et al.*, 2020]. In addition, WTe_2@CNT exhibits a reversible capacity of 592 mAh g^{-1} and 100% retention over 500 cycles at 500 mA g^{-1}, whereas pristine WTe_2 nanostars deliver high fading with a specific capacity of ~85 mAh g^{-1} over 350 cycles. These MWCNT-interconnected WTe_2 nanostars expose interlayers of active WTe_2, avoid agglomeration, maintain the structural integrity of the electrodes, and prevent volume expansion during LIB performance. TMDs have been used to a greater extent in the anode and cathode of LIBs due to their exclusive structural and electrical suitability. The 2D WS_2 with 1T metallic phase has been reported as one of the most promising electrode materials [Srinivaas *et al.*, 2019]. The 1T WS_2 NFs anode material has a rich 1T metallic phase around the active edge sites showing fast electron/ion transfer and enhanced Li capacity and cyclability. This anode exhibits charge capacity of 810 mAh g^{-1} without additional carbon support and further maintains specific capacity of 390 mAh g^{-1} at 0.2°C over 500 cycles.

9.4.2.2 *TMD-based Anode Materials for SIBs*

The high specific capacity chalcogen-based anode materials are unable to restrict huge volume expansions resulting in poor cyclic

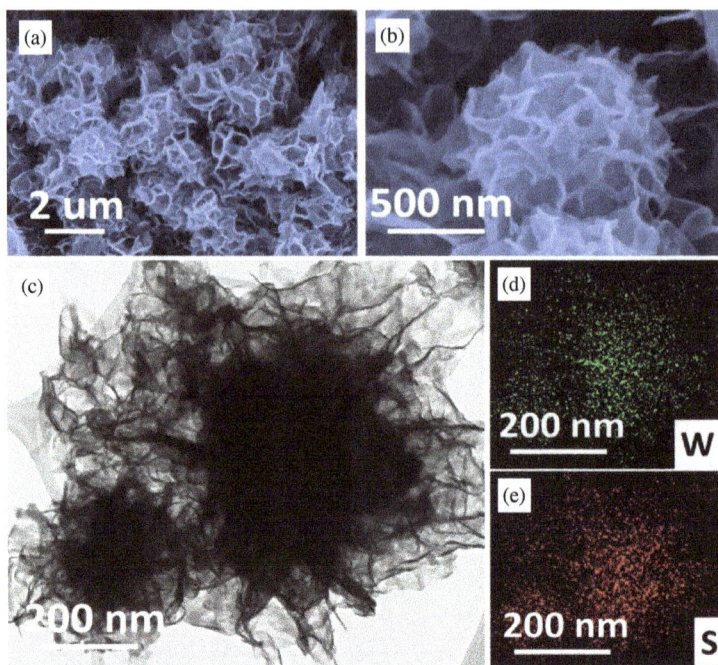

Fig 9.4 (a, b) SEM images showing morphology of few-layered WS_2 NFs at different magnifications, (c) TEM image of the synthesized WS_2 NFs, and (d, e) EDS mapping for chemical composition analysis related to tungsten (W) and sulfur (S) of the WS_2 NFs. Reprinted with permission from Srinivaas *et al.* [2019]. Copyright (2023) American Chemical Society.

performance and low coulombic efficiency for SIBs [Zhou *et al.*, 2016]. Structural modifications such as doping, size reduction, or surface coating are reported to improve the performance of anode materials for SIBs. The layered TMDs have shown unique properties and fascinating applications as anode materials for SIBs. High-quality, single-crystalline MoS_2, WS_2, WSe_2, and $MoSe_2$ nanolayers were synthesized by a NaCl template–assisted in situ CVD method. The cubic NaCl particles provide smooth surfaces to support lateral growth and restrict the thickness of TMD nanosheets. The structural stability and high crystallinity of synthesized TMD nanosheets promote the conductivity. The CV curves demonstrate two distinct reduction peaks related to the Na intercalation into WS_2 to form Na_xWS_2 and reaction of Na$_x$WS$_2$ with W nanoparticles which are intercalated in Na$_2$S matrix, respectively, as shown in Fig 9.5(a) [Zhou *et al.*, 2016]. In addition, three weak peaks emerged due to the recombination of Na$^+$ and S to Na$_2$S$_n$. Fig 9.5(b) shows charge discharge curve at 0.1 A g^{-1} and an obvious plateau is observed at ~1.8 V along with two plateaus at ~2.23 and ~2.54 V, indicating the dissociation of Na$_2$S. WS$_2$ nanosheet anode shows discharge–charge specific capacities of 574.7 and 453.2 mAh g^{-1}, respectively, with a coulombic efficiency (CE) of ~79% as shown in Fig 9.5(c). The lightening of 19 light-emitting diodes in parallel is demonstrated to show high energy density and power density after being fully charged at the 60th cycle presented in the inset of Fig 9.5(c). The rate performance of WS$_2$ nanosheets demonstrates an excellent rate capability as mentioned in Fig 9.5(d).

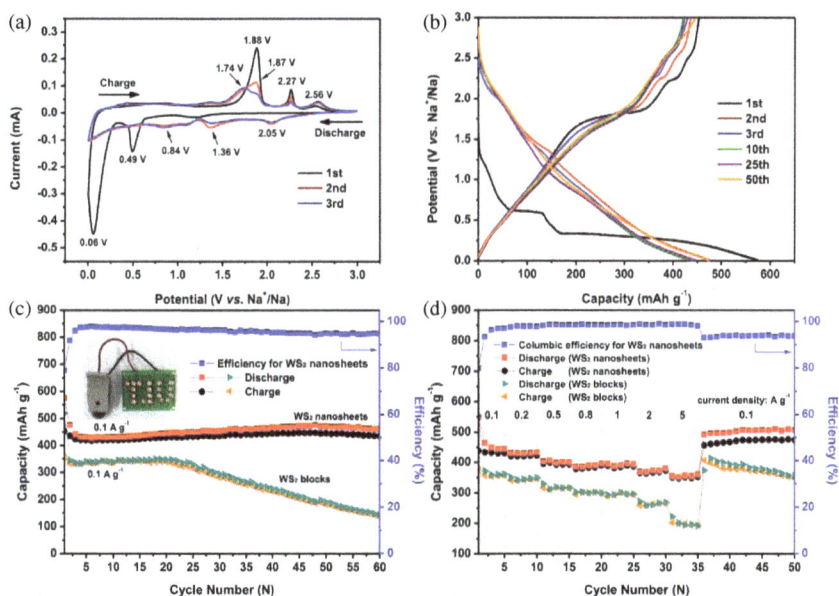

Fig 9.5   (a) CV curves of the as-obtained WS$_2$ nanosheets for the initial three cycles at a 0.1 mV s^{-1}, (b) the discharge/charge curves of above at different mentioned cycles, (c) cycling performance at 0.1 A g^{-1}, and (d) rate performances of the WS$_2$ nanosheets and WS$_2$ blocks anodes. Reprinted with permission from Zhou *et al.* [2016]. Copyright (2023) Royal Society of Chemistry.

The WS_2 nanosheets anodes for SIB exhibit a good specific capacity of 453 mAh g^{-1} at 0.1 A g^{-1} current density with improved rate capability and excellent cyclability. The exposed surface area and microstructure of WS_2 nanosheets are responsible for their superior Na-storage performance during the Na insertion/deinsertion. The TMDs show low conductivity, small interplanner spacing of MoS_2 nanosheets, and agglomeration of high surface energy nanosheets resulting in low reversible capacities and inferior cycling stability. $MoSe_2$ exhibits higher electrical conductivity and large interlayer spacing (0.64 nm) compared to MoS_2 which improves the electrochemical performance due to faster charge transfer for SIBs [Liu *et al.*, 2018]. The composite of $MoSe_2$ and electrical-conductive carbon further improve the battery performance.

The petal-like MoS_2@CNS anode with good electrical conductivity exhibits high reversible capacity (993 mAh g^{-1}) at 1 A g^{-1} over 200 cycles. However, the composite shows poor rate capability due to huge volume change (up to 300%) during electrochemical performance. The functional well-defined nanostructure formation of TMDs is also a successful technique to improve battery performance. The 3D spatial structure with few-layered TMDs, which prevents aggregation of layer nanosheets, has attracted great interest. The encapsulation-type arrangement enhances the weight ratio of the active components which increases the gravimetric energy density of SIBs. The provided gap in vicinity of core and shell buffers large volume expansions during cycling, prevents aggregation, and alleviates pulverization which improves the cycling performance. The void also facilitates path for ion conduction through electrolytes and decreases the diffusion path for electrons and Na ions. Conventionally, the formation of hollow core–shell structures is a high-cost multistep synthesis and is often failed after few electrochemical cyclings. Furthermore, the flower-like $MoSe_2$/C, with expanded (002) planes of $MoSe_2$ as anode exhibits 360 mAh g^{-1} after 350 cycles at 0.5 A g^{-1} due to enhanced movement of Na ions which promotes the Na-storage performance [Liu *et al.*, 2018].

9.4.3 *Other Nanostructured Anode Materials*

In addition, the 2D transition metal carbides/nitrides (MXene materials), Li–titanate spinel (LTO), and Si- and Ge-based materials have been introduced as good electrochemical anode material in battery applications. The anode materials such as M_2C (M = Sc, Ti, V, and Cr) have also shown good battery performance with a high capacity (>400 mAhg^{-1}). However, these materials provide a range of working potentials and they can reach the maximum capacity after structural designing and suitable composition [Wang *et al.*, 2017]. Pulverization is the main concern for high-capacity anodes which reduces the cyclability performance. LTO is introduced as very stable anode material for electrochemical performance due to very low-volume expansion during charge/discharge (<7%). However, LTO shows lower specific capacity. The group-IV (Si-, Ge-, and Sn-based) alloy-type anode materials of high theoretical capacity have a hierarchical structure which can hold more than four Li$^+$ ions, but usually have large volumetric variations (300%) during battery performance and show electrode pulverization, low capacity retention, and short cycle life. The Si and Ge nanowires (group-IV high-performance anodes) have been reported for their high electrochemical performance mentioned as >3000 mAh g^{-1} capacity with 90% efficiency for Si nanowires and >1000 mAh g^{-1} capacity with 84–96% efficiency for Ge nanowires [Ng *et al.*, 2010]. A network of nanostructured materials can prevent cracking due to better mechanical properties during the electrochemical performance. Different types of materials are used as anodes for LIBs which are discussed here.

9.4.3.1 *Other Nanostructured Anode Materials for LIBs*

Both carbon and other nanomaterials show strength as anode materials to intercalate large amounts of Li ions and by alloy formation in group-IV nanowires with two Li-uptake mechanisms. Simultaneously, it improves the ion-storage performance of LIBs [Ng *et al.*, 2010].

The Sn nanowire covering with conducting CNT network (SnNW@ CNT) as an anode material, which forms crystalline Sn with preferred growth orientation of ⟨001⟩ and theoretical capacity of 994 mAhg^{-1}, has shown remarkable specific capacitance for LIBs [Li *et al.*, 2007]. Hybrid nanomaterial anodes such as SnO_2/In_2O_3 and SnO_2/Sn nanoclusters have been reported experimentally with improved efficiency for LIBs [Kim *et al.*, 2007]. The spinel $Li_4Ti_5O_{12}$ (LTO) anode material performs poor Li-storage capacity retention and rate capability due to a low electrical conductivity ($<10^{-13}$ S cm^{-1}) of the material [Vikram Babu *et al.*, 2018]. In addition, RGO additives in composite with LTO electrode material have improved the electrical conductivity and shortened the ion-diffusion pathways, resulting in improved reversible capacity (124 mAhg^{-1} at a 20°C) and rate performance of LTO [Sun *et al.*, 2015]. The charge-transfer resistance (R_{ct}) in the Nyquist plots for LTO-RGO and LTO-C composites shows the value of 230 and 450 Ω, respectively. It is concluded that the effective Li$^+$-ion and electron transportation of LTO-RGO has been increased by adding graphene during charge–discharge processes. The high-capacity Si- and Sn-based anode materials suffer from pulverization of the particles and poor electrical conductivity. The conducting graphene-based anode materials, offering efficient electrical-conducting channels, improve reversible capacity and cyclability with noticeable specific capacity (>600 mAh g^{-1}) and low lithiation voltage of 0.29 V vs Li/Li$^+$. The graphene-based coatings on other semiconducting anode materials such as MnO_2, MoS_2, and phosphorus also result in a higher rate capability and an outstanding cyclability owing to an interconnected conductive arrangement to utilize maximum active anode particles. Fe_3O_4–Ti_3C_2 hybrid materials show that the 2D-layered nanostructured Ti_3C_2 holds the volume change and provides path for speedy diffusion of Li$^+$ during charge–discharge performance. In addition, the pulverization of the Fe_3O_4 nanoparticles improves the contact of Fe_3O_4 nanoparticles with Ti_3C_2 MXene during cycles which favors charge-transfer kinetics and leads to high electrochemical performance. The black phosphorus and its alloying (2D-nanolayered phosphorene) anode materials with high theoretical capacity, which provide ultrahigh Li^+ ion mobility, show large volume changes and suffer rapid capacity loss during cycling. The hexagonal boron nitride nanosheets (BNNs) as a capping agent of phosphorene was studied through modeling and showed that improvement in the structural stability reduces the volume change during electrochemical performance and can be used to sort-out poor electrochemical performance of phosphorene while maintaining its high specific capacity.

9.4.3.2 *Other Nanostructured Anode Materials for SIBs*

The safe, low-cost, nontoxic titanium-based anode materials such as titanium dioxide are used as promising candidates for Na insertion with low redox potentials ($Ti^{3+}/^{4+}$) due to shortening of the Na ions diffusion path [Wang *et al.*, 2017]. The low-cost, safe, and stable Ti-based compounds, including polymorphs TiO_2, $Na_2Ti_nO_{2n+1}$ ($2 \leq n \leq 9$), and its derivates with low redox potentials, are suitable anode materials for SIBs [Wang *et al.*, 2013, 2015]. The low lattice strain and superior structural stability of these compounds are achieved due to 2D and 3D frameworks of TiO_6 octahedra during Na insertion/extraction [Wang *et al.*, 2013, 2017]. These Ti-based anode materials show good cycle stability and acceptable voltage range. On the contrary, they face low specific capacity (<120 mAh g^{-1}) and low coulombic efficiency [Wang *et al.*, 2015]. The low-cost organic materials as an anode of SIBs are also appropriate candidates due to tunable voltage range and potential multielectron reactions. However, they show poor cycling performance and low initial coulombic efficiency, which require further improvement. The oxides, sulfides, or selenides of groups XIV and XV elements, (chalcogen-based compounds, e.g., SnO, SnO_2, SnS), react with Na via a conversion reaction and form alloys [Wang *et al.*, 2017]. The alloy anode materials which show the highest theoretical capacity react with Na to form Na–metal alloys during the charge–discharge performance with a huge volume expansion for SIBs. The chalcogen-based compounds as anode materials work on combination of conversion and alloying reaction to store Na during SIBs performance.

9.5 Summary and Future Perspectives

The anode materials show dendrites formation, delamination of foil current collectors, and volume and structural changes, which lead to fast capacity decay resulting in low electrochemical performance during cycling. In addition, batteries fail to perform at extreme environmental conditions and create crucial safety concerns. These nanostructured anode materials such as graphene, graphene analogs, and chalcogenides 2D materials show corrosion-protective properties, thermal safety, mechanical stability, good cyclability, high volumetric and gravimetric capacities, prevent dendrite formation and pulverization, and are found to be potential candidates to handle environmental sustainability and safety issues in battery applications. Still, the higher capacity chalcogen and alloy compound anode materials suffer with uncontrollable volume change which is a key challenge for battery applications. The nanostructured engineering, conductive materials coating, and heteroatom doping are well-established techniques to further improve the anode material properties for both LIBs and SIBs. The performance of anode materials depends on structural defects, impurities, and graphitization, which result in increased surface area, larger interlayer distance, and reduced diameter. In addition, anode materials made of aerogels, CNT foam, and graphene-based fiber cloth are used for elastic stretchable anode material for self-supported LIBs. The nanostructured anode materials used for LIBs are not appropriate for Na-ion intercalation in SIBs and show their large volume expansion, low initial coulombic efficiency, and moderate rate performance due to larger size of Na ion and unfavorable thermodynamics. Various types of nanostructured anode materials are reported for SIBs. They are soft and hard amorphous carbons, chalcogen-based, alloy materials, titanium-based, and organic materials. The defects are created in soft and hard carbon by heteroatom doping, for example, nitrogen, sulfur, and phosphorous, pyrolysis to get controlling microstructure, and more defect sites of solid-phase precursors. The elevated specific surface area and good electronic conductivity generated by defects provide higher reversible capacity and higher sodiation potential in SIBs. The formation of ultrathin nanosheets of 2D TMD with atomic-scale thickness as anode exhibits outstanding surface area, mechanical and electrical properties which generate superior electrochemical properties for both LIBs and SIBs due to its small band gap and polymorphic metastable metallic phase. In addition, the increase in interplanner spacing of TMDs can help promote carrier-diffusion kinetics and superior electronic conductivities during Li- and Na-ion transportation. Further, the 3D spatial structure with few-layered TMDs, which prevents aggregation of layer nanosheets and cracking due to better mechanical properties during the electrochemical performance, has attracted significant interest. The graphene-based coatings on other semiconducting anode materials form an interconnected conductive framework, resulting in greater rate capability and outstanding cyclability due to maximum active anode particle utilization.

References

Ali, G., Oh, S. H., Kim, S. Y., Kim, J. Y., Cho, B. W. & Chung, K. Y. (2015). An open-framework iron fluoride and reduced graphene oxide nanocomposite as a high-capacity cathode material for Na-ion batteries. *J. Mater. Chem. A* 3(19), pp. 10258–10266.

Bachtold, A., Strunk, C., Salvetat, J. P., Bonard, J. M., Forró, L., Nussbaumer, T. & Schönenberger, C. (1999). Aharonov-bohm oscillations in carbon nanotubes. *Nature* 397(6721), pp. 673–675.

Canal-Rodríguez, M., Arenillas, A., Menéndez, J. A., Beneroso, D. & Rey-Raap, N. (2018). Carbon xerogels graphitized by microwave heating as anode materials in lithium-ion batteries. *Carbon* 137, pp. 384–394.

Fang, S., Bresser, D. & Passerini, S. (2020). Transition metal oxide anodes for electrochemical energy storage in lithium-and sodium-ion batteries. *Adv. Energy Mater.* 10(1), p. 1902485.

Fu, L. J., Liu, H., Li, C., Wu, Y. P., Rahm, E., Holze, R. & Wu, H. Q. (2006). Surface modifications of electrode materials for lithium ion batteries. *Solid State Sci.* 8(2), pp. 113–128.

Ge, P. & Fouletier, M. (1988). Electrochemical intercalation of sodium in graphite. *Solid State Ion.* 28–30(Part 2), pp. 1172–1175.

Hassoun, J., Bonaccorso, F., Agostini, M., Angelucci, M., Betti, M. G., Cingolani, R., Gemmi, M., Mariani, C., Panero, S., Pellegrini, V. & Scrosati, B. (2014). An advanced lithium-ion battery based on a graphene anode and a lithium iron phosphate cathode. *Nano Lett.* 14(8), pp. 4901–4906.

Kim, D. W., Hwang, I. S., Kwon, S. J., Kang, H. Y., Park, K. S., Choi, Y. J., Choi, K. J. & Park, J. G. (2007). Highly conductive coaxial SnO_2-In_2O_3 heterostructured nanowires for Li Ion battery electrodes. *Nano Lett.* 7(10), pp. 3041–3045.

Landi, B. J., DiLeo, R. A., Schauerman, C. M., D. Cress, C., J. Ganter, M. & P. Raffaelle, R. (2009). Multi-walled carbon nanotube paper anodes for u lithium ion batteries. *J. Nanosci. Nanotechnol.* 9(6), pp. 3406–3410.

Li, Y., Hu, Y. S., Qi, X., Rong, X., Li, H., Huang, X. & Chen, L. (2016). Advanced sodium-ion batteries using superior low cost pyrolyzed anthracite anode: Towards practical applications. *Energy Stor. Mater.* 5, pp. 191–197.

Li, R., Sun, X., Zhou, X., Cai, M. & Sun, X. (2007). Aligned heterostructures of single-crystalline tin nanowires encapsulated in amorphous carbon nanotubes. *J. Phys. Chem. C* 111(26), pp. 9130–9135.

Li, Y., Wang, R., Guo, Z., Xiao, Z., Wang, H., Luo, X. & Zhang, H. (2019). Emerging two-dimensional noncarbon nanomaterials for flexible lithium-ion batteries: Opportunities and challenges. *J. Mater. Chem. A* 7(44), pp. 25227–25246.

Liu, H., Guo, H., Liu, B., Liang, M., Lv, Z., Adair, K. R. & Sun, X. (2018). Few-Layer $MoSe_2$ nanosheets with expanded (002) planes confined in hollow carbon nanospheres for ultrahigh-performance Na-Ion batteries. *Adv. Funct. Mater.* 28(19), pp. 1–9.

Lv, X., Deng, J., Wang, J., Zhong, J. & Sun, X. (2015). Carbon-coated α-Fe_2O_3 nanostructures for efficient anode of Li-ion battery. *J. Mater. Chem. A* 3(9), pp. 5183–5188.

Nan, D., Huang, Z. H., Lv, R., Yang, L., Wang, J. G., Shen, W., Lin, Y., Yu, X., Ye, L., Sun, H. & Kang, F. (2014). Nitrogen-enriched electrospun porous carbon nanofiber networks as high-performance free-standing electrode materials. *J. Mater. Chem. A* 2(46), pp. 19678–19684.

Ng, M. F., Zheng, J. & Wu, P. (2010). Evaluation of sn nanowire encapsulated carbon nanotube for a li-ion battery anode by dft calculations. *J. Phys. Chem. C* 114(18), pp. 8542–8545.

Paek, S. M., Yoo, E. J. & Honma, I. (2009). Enhanced cyclic performance and lithium storage capacity of SnO_2/graphene nanoporous electrodes with three-dimensionally delaminated flexible structure. *Nano Lett.* 9(1), pp. 72–75.

Srinivaas, M., Wu, C. Y., Duh, J. G., Hu, Y. C. & Wu, J. M. (2020). Multi-walled carbon-nanotube-decorated tungsten ditelluride nanostars as anode material for lithium-ion batteries. *Nanotechnology* 31(3), p. 035406.

Srinivaas, M., Wu, C. Y., Duh, J. G. & Wu, J. M. (2019). Highly Rich 1T Metallic Phase of Few-Layered WS_2 Nanoflowers for Enhanced Storage of Lithium-Ion Batteries. *ACS Sustain. Chem. Eng.* 7(12), pp. 10363–10370.

Sun, J., Lee, H. W., Pasta, M., Yuan, H., Zheng, G., Sun, Y., Li, Y. & Cui, Y. (2015). A phosphorene-graphene hybrid material as a high-capacity anode for sodium-ion batteries. *Nat. Nanotechnol.* 10(11), pp. 980–985.

Sun, X., Radovanovic, P. V. & Cui, B. (2015). Advances in spinel Li4Ti5O$_{12}$ anode materials for lithium-ion batteries. *New J. Chem.* 39(1), pp. 38–63.

Vikram Babu, B., Vijaya Babu, K., Tewodros Aregai, G., Seeta Devi, L., Madhavi Latha, B., Sushma Reddi, M., Samatha, K. & Veeraiah, V. (2018). Structural and electrical properties of $Li4Ti_5O_{12}$ anode material for lithium-ion batteries. *Results Phys.* 9, pp. 284–289.

Wan, J., Shen, F., Luo, W., Zhou, L., Dai, J., Han, X., Bao, W., Xu, Y., Panagiotopoulos, J., Fan, X., Urban, D., Nie, A., Shahbazian-Yassar, R. & Hu, L. (2016). In situ transmission electron microscopy observation of sodiation-desodiation in a long cycle, high-capacity reduced graphene oxide sodium-ion battery anode. *Chem. Mater.* 28(18), pp. 6528–6535.

Wang, G., Shen, X., Yao, J. & Park, J. (2009). Graphene nanosheets for enhanced lithium storage in lithium ion batteries. *Carbon* 47(8), pp. 2049–2053.

Wang, T., Shi, S., Li, Y., Zhao, M., Chang, X., Wu, D., Wang, H., Peng, L., Wang, P. & Yang, G. (2016). Study of microstructure change of carbon nanofibers as binder-free anode for high-performance lithium-ion batteries. *ACS Appl. Mater. Interfaces* 8(48), pp. 33091–33101.

Wang, M. S., Song, W. L. & Fan, L. Z. (2015). Three-dimensional interconnected network of graphene-wrapped silicon/carbon nanofiber hybrids for binder-free anodes in lithium-ion batteries. *Chem. Electro. Chem.* 2(11), pp. 1699–1706.

Wang, Y., Xiao, R., Hu, Y. S., Avdeev, M. & Chen, L. (2015). P_2-$Na_{0.6}$ [$Cr_{0.6}$ $Ti_{0.4}$]O_2 cation-disordered electrode for high-rate symmetric rechargeable sodium-ion batteries. *Nat. Commun.* 6, pp. 1–9.

Wang, Y., Yu, X., Xu, S., Bai, J., Xiao, R., Hu, Y. S., Li, H., Yang, X. Q., Chen, L. & Huang, X. (2013). A zero-strain layered metal oxide as the negative electrode for long-life sodium-ion batteries. *Nat. Commun.* 4, pp. 1–8.

Wang, Q., Zhao, C., Lu, Y., Li, Y., Zheng, Y., Qi, Y., Rong, X., Jiang, L., Qi, X., Shao, Y., Pan, D., Li, B., Hu, Y. S. & Chen, L. (2017). Advanced nanostructured anode materials for sodium-ion batteries. *Small* 13(42), pp. 1–32.

Wei, W., Yang, S., Zhou, H., Lieberwirth, I., Feng, X. & Müllen, K. (2013). 3D graphene foams cross-linked with pre-encapsulated Fe3O 4 nanospheres for enhanced lithium storage. *Adv Mater.* 25(21), pp. 2909–2914.

Yazami, R. & Touzain, P. (1983). A reversible graphite-lithium negative electrode for electrochemical generators. *J. Power Sources* 9(3), pp. 365–371.

Yoo, E. J., Kim, J., Hosono, E., Zhou, H. S., Kudo, T. & Honma, I. (2008). Large reversible Li storage of graphene nanosheet families for use in rechargeable lithium ion batteries. *Nano Lett.* 8(8), pp. 2277–2282.

Yoon, S. H., Park, C. W., Yang, H., Korai, Y., Mochida, I., Baker, R. T. K. & Rodriguez, N. M. (2004). Novel carbon nanofibers of high graphitization as anodic materials for lithium ion secondary batteries. *Carbon* 42(1), pp. 21–32.

Yuan, X., Chao, Y. J., Ma, Z. F. & Deng, X. (2007). Preparation and characterization of carbon xerogel (CX) and CX-SiO composite as anode material for lithium-ion battery. *Electrochem. Commun.* 9(10), pp. 2591–2595.

Zhang, W. M., Wu, X. L., Hu, J. S., Guo, Y. G. & Wan, L. J. (2008). Carbon coated Fe3O4 nanospindles as a superior anode material for lithium-ion batteries. *Adv. Funct. Mater.* 18(24), pp. 3941–3946.

Zhou, J., Qin, J., Guo, L., Zhao, N., Shi, C., Liu, E. Z., He, F., Ma, L., Li, J. & He, C. (2016). Scalable synthesis of high-quality transition metal dichalcogenide nanosheets and their application as sodium-ion battery anodes. *J. Mater. Chem. A* 4(44), pp. 17370–17380.

Index